T0142808

Multi-finger Haptic Interaction

Springer Series on Touch and Haptic Systems

For further volumes:
www.springer.com/series/8786

Ignacio Galiana · Manuel Ferre
Editors

Multi-finger Haptic Interaction

Editors
Ignacio Galiana
Centre for Automation and Robotics
UPM-CSIC
Universidad Politécnica de Madrid
Madrid, Spain

Manuel Ferre
Centre for Automation and Robotics
UPM-CSIC
Universidad Politécnica de Madrid
Madrid, Spain

ISSN 2192-2977
ISSN 2192-2985 (electronic)
Springer Series on Touch and Haptic Systems
ISBN 978-1-4471-5852-3
ISBN 978-1-4471-5204-0 (eBook)
DOI 10.1007/978-1-4471-5204-0
Springer London Heidelberg New York Dordrecht

Printed on acid-free paper

Springer is part of Springer Science+Business Media (www.springer.com)

Series Editors' Foreword

This is the seventh volume of the "Springer Series on Touch and Haptic Systems", which is published in a collaboration between **Springer** and the **EuroHaptics Society**.

Multi-finger Haptic Interaction is focused on characterizing and reproducing multi-finger object grasping actions. The growing importance of this endeavour is demonstrated by potential robotic applications such as object transport and manipulation tasks as well as growing recognition of the need to effect handover of objects between robots and humans where the multi-digit nature of the individual and combined grasps is critical to success.

The contributions in this book include an overview of the available technologies for multi-finger haptic interaction, a description of mechanical design and control issues, development of methods to allow multiple contact points when interacting with virtual scenarios, specification of techniques to allow bimanual manipulation and reports on human perception studies. Overall these studies offer a key to understanding how multi-finger haptic feedback improves the realism and perception for a variety of different applications.

The editors of *Multi-finger Haptic Interaction* are Ignacio Galiana and Manuel Ferre who have extensive experience in the field. Galiana is research engineer and PhD candidate at the Centre for Automation and Robotics UPM-CSIC. Ferre is professor and vice-director of the Centre for Automation and Robotics UPM-CSIC. He is also one of the Editors of the Springer Series on Touch and Haptic Systems. The book arose out of Galiana and Ferre's joint interests in developing multi-finger haptic devices and methods for enhanced immersion in simulated environments and remote handling and an international conference workshop they organised at the Eurohaptics meeting in Finland in 2011.

March 2013

Marc Ernst
Alan Wing

Contents

Contributors

Katsuhito Akahane Tokyo Institute of Technology, Midori, Yokohama, Japan

Jorge Barrio Centre for Automation and Robotics UPM-CSIC, Universidad Politécnica de Madrid, Madrid, Spain

Alastair Barrow Imperial College, London, UK

Jose M. Breñosa Centre for Automation and Robotics UPM-CSIC, Universidad Politécnica de Madrid, Madrid, Spain

Gabriel Cirio IRISA/INRIA, Rennes, France

Vytautas Daniulaitis UAB Aksonas, Kaunas, Lithuania

Takahiro Endo Gifu University, Gifu, Japan

Manuel Ferre Centre for Automation and Robotics UPM-CSIC, Universidad Politécnica de Madrid, Madrid, Spain

Ignacio Galiana Centre for Automation and Robotics UPM-CSIC, Universidad Politécnica de Madrid, Madrid, Spain

William Harwin School of Systems Engineering, University of Reading, Reading, UK

Jungwoo Hyun Tokyo Institute of Technology, Midori, Yokohama, Japan

Yasuhiko Ishigure Marutomi Seiko Co., Ltd., Seki-shi, Gifu, Japan

Haruhisa Kawasaki Gifu University, Gifu, Japan

Itsuo Kumazawa Tokyo Institute of Technology, Midori, Yokohama, Japan

Anatole Lécuyer IRISA/INRIA, Rennes, France

Maud Marchal IRISA/INRIA/INSA, Rennes, France

Tetuya Mouri Gifu University, Gifu, Japan

Makoto Sato Tokyo Institute of Technology, Midori, Yokohama, Japan

Anthony Talvas IRISA/INRIA/INSA, Rennes, France

Chapter 1
Introduction

Ignacio Galiana and Manuel Ferre

Abstract Real-life interactions involve dexterous manipulation with multiple fingers and palm; however, most commercially available devices allow haptic interaction with only one contact point. In this book, we provide a panoramic state of the art of current developments in the fields of Multi-finger Haptic Technologies, multi-contact point simulation algorithms, interaction techniques and implications in human perception. This introductory chapter provides motivation for developing haptic technologies and methods with multiple fingers followed by a detailed description of chapters composing this book issue.

Multi-finger Haptic methods provide a realistic and natural human-machine interface that enhances immersion when interacting with simulated or remote environments. This issue provides a panoramic state of the art of methods for multi-finger haptic interaction. Implications of adding multiple-fingers to haptic perception are also analyzed. Topics covered in this issue include: design and control of advanced haptic devices; multi-contact point simulation algorithms, interaction techniques and implications in human perception when interacting with multiple fingers. These multi-disciplinary results are integrated into applications such as medical simulators for training manual skills, simulators for virtual prototyping, precise manipulations in remote environments, and control of robotic hands.

Most commercial devices currently available allow haptic interaction with only one contact point, which may be sufficient for some exploration or palpation tasks; however, multi-contact point applications allow performing more advanced and realistic object manipulation such as grasping or assembling. The realism of multi-contact point applications is significantly higher than the classical one-contact point, since the ways we are used to interacting in our daily lives (using a toothbrush, opening doors, using cell-phones, playing basketball, etc.), or in professional situations

I. Galiana (✉) · M. Ferre
Centre for Automation and Robotics UPM-CSIC, Universidad Politécnica de Madrid,
C/ José Gutiérrez Abascal 2, 28006 Madrid, Spain
e-mail: ignacio.galiana@upm.es

M. Ferre
e-mail: m.ferre@upm.es

I. Galiana, M. Ferre (eds.), *Multi-finger Haptic Interaction*,
Springer Series on Touch and Haptic Systems, DOI 10.1007/978-1-4471-5204-0_1,

(medical interventions, manufacturing processes, etc.) typically involve dexterous manipulations with multiple fingers and palm for both hands.

Considering multiple fingers for haptic applications opens new paradigms such as cooperative and bimanual manipulation. In general, the current one contact point devices, scenarios and rendering methods cannot be directly scaled to allow multi-finger haptic interaction. In order to provide a realistic interaction, advances in the following areas are required: (i) new haptic rendering methods that compute more complex manipulation scenarios in real time, (ii) mechanical devices that allow force-feedback to multiple fingers avoiding mechanical collisions, as well as (iii) methods that amplify the interaction workspace to a whole virtual world or remote scenario.

In this book, an overview of the available technologies for multi-finger haptic interaction, mechanical design and control are described, methods to allow multiple contact points when interacting with virtual scenarios, techniques to allow bimanual manipulation and human perception studies are also carried out and described in parallel to these technological developments. Some key points of these studies are related to the assessment of how multi-finger haptic feedback improves the realism and perception for different applications.

Contributions presented in this issue are oriented to researchers who are involved in haptic device design, rendering methods and perception studies as well as readers with different backgrounds who are interested in applying multi-finger haptic technologies and methods to their field of interest or who want to familiarize themselves with the above described concepts.

Five works which show a nice view of the potential applications that can be developed with these systems and their complexity are included in this book. All these works are focused on haptic interaction by providing multiple contact points to the user.

In Chap. 2, haptic rendering methods for *multi-finger grasps in a dynamic environment* based on object models and the friction cone algorithm are analyzed. A multi-threaded haptic render that incorporates these models to allow interactions with up to all digits of both hands is described in this chapter. These methods have been successfully implemented and applied in several applications developed at University of Reading in the UK.

3D Interaction Techniques for Bimanual Haptics in Virtual Environments are described in Chap. 3. Bimanual haptic manipulation is typically limited by the current workspace of haptic devices. This chapter describes the double bubble technique, which allows workspace extension for bimanual haptic interaction with virtual environments. Grasping techniques in order to improve manipulation of objects for bimanual manipulations are studied in this chapter and also evaluated with user studies. These techniques were developed at IRISA/INRIA in France.

Chapter 4 describes a *Modular Multi-finger Haptic Device: Mechanical Design, Controller and Applications*. The designed device is called MasterFinger, this scalable device can be configured to allow 1, 2 or 3 fingers manipulation. Mechanical design and control architecture are described in detail. Additionally, applications of this device to simulate medical rehabilitation procedures, manipulate fragile objects and to perform collaborative manipulation between two users are shown. These applications highlight the performance of this device and the importance of haptics to improve user perception. This haptic interface has been developed at the Centre for Automation and Robotics UPM-CSIC in Spain

Chapter 5 describes *HIRO: Multi-fingered Haptic Interface Robot and Its Medical Application Systems*. This haptic device has been developed at University of Gifu in Japan. Researchers at U. of Gifu have been working in this system for more than a decade and they are currently commercializing the HIRO device. The presented device provides high-precision three-directional force at the five human fingertips; moreover, this device is connected to a robotic arm in order to provide a large workspace. Detailed description of the mechanical and control design is included, moreover, several applications for medical training and rehabilitation are described by authors.

In Chap. 6, *Two-Handed, Multi-finger, String-Based, Haptic Interface Spidar-8*, authors designed a device that is able to provide accurate force feedback to up to eight fingertips by using strings to transmit force. This design minimizes the weight and inertia of the Spidar device. Researchers at Tokyo Institute of Technology also developed an augmented reality user interface in which the user's real hands are visually merged into the virtual world and displayed to the user, significantly increasing the realism of their applications. In this chapter, mechanical and control design as well as successful applications to user studies in dexterous manipulation and virtual prototyping carried out at Tokyo Institute of Technology are described.

Chapter 2
Multi-finger Grasps in a Dynamic Environment

William Harwin and Alastair Barrow

Abstract Most current state-of-the-art haptic devices render only a single force, however almost all human grasps are characterised by multiple forces and torques applied by the fingers and palms of the hand to the object. In this chapter we will begin by considering the different types of grasp and then consider the physics of rigid objects that will be needed for correct haptic rendering. We then describe an algorithm to represent the forces associated with grasp in a natural manner. The power of the algorithm is that it considers only the capabilities of the haptic device and requires no model of the hand, thus applies to most practical grasp types. The technique is sufficiently general that it would also apply to multi-hand interactions, and hence to collaborative interactions where several people interact with the same rigid object. Key concepts in friction and rigid body dynamics are discussed and applied to the problem of rendering multiple forces to allow the person to choose their grasp on a virtual object and perceive the resulting movement via the forces in a natural way. The algorithm also generalises well to support computation of multi-body physics

2.1 Introduction

Our ability to manipulate our environment underlies our intelligence, and the role of hands in this manipulation is self evident. The machinery behind the hand, both mechanical (muscular skeletal) and computational (neuronal) has been an inspiration for research and design. Hooks and artificial hands are the currency of upper-limb prosthetics and an early example was the iron hand made for the soldier Götz von Berlichingen in the 15th Century. In the early part of the 20th Century the value of remote handling was realised with the first master-slave telemanipulators [41].

W. Harwin (✉)
School of Systems Engineering, University of Reading, Reading RG6 6AY, UK
e-mail: w.s.harwin@reading.ac.uk

A. Barrow
Imperial College, London, UK
e-mail: a.barrow@imperial.ac.uk

I. Galiana, M. Ferre (eds.), *Multi-finger Haptic Interaction*,
Springer Series on Touch and Haptic Systems, DOI 10.1007/978-1-4471-5204-0_2,

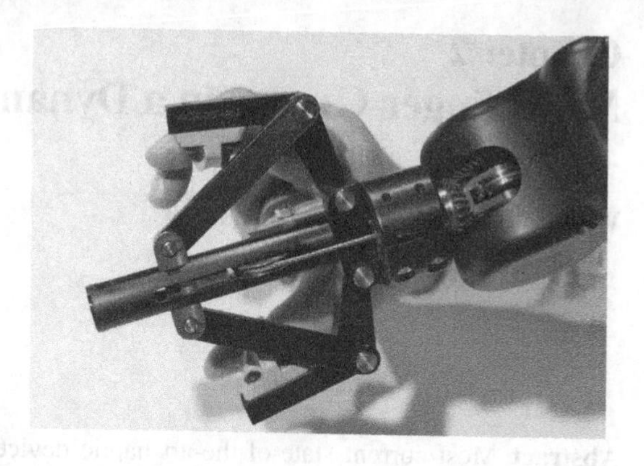

Fig. 2.1 Remote handling master hand interface mechanism. Courtesy of Oxford Technologies Ltd. (www.oxfordtechnologies.co.uk)

The early master-slave telemanipulators were directly coupled so the forces encountered in the environment were reflected into the hands. Modern telemanipulators are now coupled though an information channel but the value of the force reflection is still realised in modern implementations to give the operator greater control (Fig. 2.1). These ideas were combined with the concepts of simulated environments and virtual realities and led to the idea of telepresence. The interest in haptics then emerged as a mechanism to facilitate our interaction with computers and visual information [6, 7], but with a much devalued concept of manipulation. Multi-finger concepts provide a means to realise manipulation in virtual and simulated environments.

Most modern haptic devices associate interactions between the person's hand and a virtual environment which can be optionally colocated (an example of co-location is shown in Fig. 2.14). The high cost of haptic device technology has resulted in most of these interactions occurring via a single point of contact which could be a thimble as in the Phantom,[1] or via a tool surrogate. Surrogates include primitive shapes such as the spheres used by the Falcon,[2] HapticMaster,[3] and Omega.3;[4] or tools such as pens/styli, dental and laproscopic instruments, catheters and needles. Although interaction with surrogate tools may simplify hand-world interactions, the problem is that the tool may need to be changed for each context. So, for example, a training workstation used to teach dentistry skills may use a surrogate of the dental hand piece to remove filling or decayed tooth (caries), but this tool surrogate will need to be exchanged for a thinner and lighter stylus to represent a probe for diagnosing decay, a periodontal probe to measure pocket depth or a periodontal scaler for removing calculus [12, 44].

Allowing force or force+torque (wrench) interactions to happen at the level of the individual contacts between the person and the virtual object allows more flexibility in the individual applications, in particular it eliminates the need for applica-

[1] Sensable.com, USA.

[2] Novint, USA.

[3] Moog FCS, The Netherlands.

[4] ForceDimension, Switzerland.

tion dependent surrogate objects. There are two reasons that this is a less favoured approach, firstly cost since it increases the number of degrees of freedom needed, and second the complexity, since problems emerge in terms of reducing the effective workspace over a simpler device and the unintended self collisions between the linkages of the haptic device, or between the device and the person's fingers, palm etc.

However there is a growing interest multi-finger and multi-contact haptics. Work has been done on kinematic design of multi-finger interactions [4, 48], and on providing multi-contact haptics in large workspaces [5, 35, 50]. Work reported elsewhere in this book includes the Spidar8 and Spidar-hand by Kumazawa and Sato [21]. The latter project succeeded in providing 8 contact points to 4 fingers on each hand using a total of 24 cables (3 per finger). The HIRO series of robot hands [15] is also reported elsewhere in this book and provides an innovative method to deliver forces to the fingers and thumbs of each hand. Reach and grasp has been studied also in terms of stroke rehabilitation and work by Loureiro et al. [27, 28] has shown that there are benefits in retraining a reach-grasp-transport-release cycle for people with upper limb hemiplegia. The Gentle/G system used for the study relied on the HapticMaster admittance controlled haptic device for the gross movements, and a custom built three-axis admittance controlled hand exoskeleton for the grasp-release cycles.

There has been less work on assessment of multi-finger contacts, but McKnight [29] has shown that the addition of multi-finger haptic feedback leads to better positioning accuracy in 5 degrees of freedom.

2.2 Grasp Analysis

Although the algorithm now reported does not require knowledge of hand anatomy, physiology or motor control, these concepts form an important backbone for the design and evaluation of any haptic device so are discussed here in outline.

There are many taxonomies for hand grasps and work has been done in areas such as prosthetics [26], psychology [16], occupational and physical therapy [45], ergonomics, robotics [14], haptics [8], accessibility education and employment.

One classification is shown in Fig. 2.2 (US access board) which shows a broad classification mechanism into pinch, lateral pinch, power, etc. However if you consider the palm as a reference frame it is possible to identify 16–20 DoF as described in Table 2.1. Degrees of freedom are often linked, or the range of movement is only a few degrees hence the uncertainty in defining the independent degrees of freedom. However as shown in Fig. 2.3, even the simple act of holding a pencil or stylus can elicit a range of acceptable grasps. More recently in-hand actions have become prominent due to the rise of mobile phone texting, and a typical grasp would lay the keyboard of the phone across the fingers to allow texting with the thumb as shown in Fig. 2.4.

We consider grasp taxonomy in terms of the physics needed for a stable grasp, that is a set of contact points or areas between the person's hand and the object

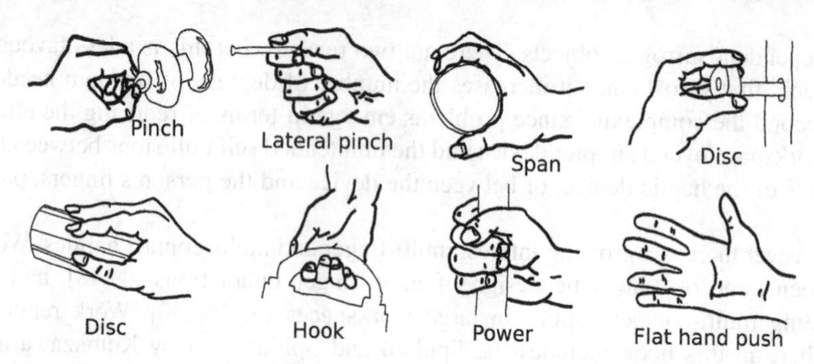

Fig. 2.2 Some typical Grip configurations (Reprinted from monograph by Edward Steinfeld, Hands-on Architecture [42])

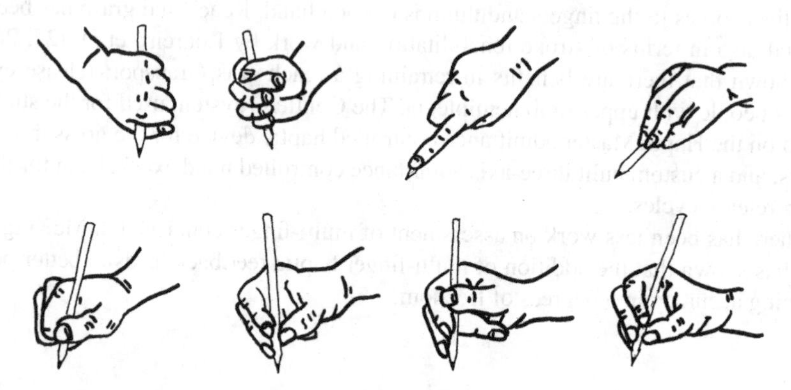

Fig. 2.3 The classic stylus or pencil grasp is called a 'tripod grasp'. In practice there is a large variation of stylus grasps. Adapted from [40]

Table 2.1 Principal degrees of freedom within the hand

Joint	DoF
Metacarpal proximal phalengeal (fingers)	8
Proximal to mid phalengeal (finger)	4
Mid to distal phalengeal (finger)	4
Carpal to metacarpal (thumb)	2
Metacarpal to proximal phalengeal (thumb)	1
Proximal to distal (thumb)	1

that allow the object to be accelerated without danger of losing control over the trajectory of the object. An approximate hierarchy of complexity is

- Single point of contact: Static stability requires object mass to lie below point of contact. Example would be a finger hooked to hold a cup or mug by its handle.

Fig. 2.4 Typical grasp while texting with the thumb

- Single area of contact: Stability requires the object centre of mass to rise if the object is perturbed. Example would be holding a plate on the palm.
- Multiple points of contact on the finger(s) and thumb. Stability is discussed below. Examples are pinch grips, and stylus grips.
- Multiple points or areas of contact that could include finger(s), thumb and palm. All grasps could be considered in terms of areas of contact that might change over the duration of the grasp. The definition would now include manipulating a mobile phone to do texting, as well as what are some times known as power grasps, where the hand encloses the object, for example the handle of a hammer, or a can of soft-drink.

A further requirement is that the contacting fingers have the span to reach to the considered contact points as well as the capacity to exert the necessary force. This is sometimes defined as grasp isotropy [14].

Assuming that grasp isotropy can be achieved for a particular individual the requirement for a stable grasp is then force closure. Force closure is the condition where the net forces on the object from the fingers, gravity, any other objects or constraints, and due to any accelerations must be zero. The robustness of the grasp can then be considered as the perturbation of all the forces applied to the object that still maintain the object within the person's grasp. This condition will be used as the basis for the multi-finger haptic rendering described below.

A second, and related condition required for a stable grasp is torque closure. Point forces applied via the fingers to the object result in a torque around the object's centre of gravity. Where a contact area is involved in the grasp then an additional torque is applied as being the sum of the integral of the pressure taken as a moment around the centre of force. The discussion below will only consider the torques due to point forces, and will assume that the torque due to the pressure distribution over the contact area can be ignored, however extending this algorithm to include the latter should not be difficult.

If a contact area is being considered then the point equivalent force $\underline{F}$ can be calculated as

$$\underline{F} = \int_A P(\underline{r})\, d\underline{a}$$

where $\underline{P}$ is the pressure field at point $\underline{r}$ integrated over contact area A. The location of this force is the centroid of the pressure distribution. This force must be combined with a torque calculated as

$$\underline{T} = \int_A P(\underline{r})\, \underline{r} \times d\underline{a}$$

The concepts of force closure and torque closure can be used to define a stable grasp. The concept is modified slightly in haptic rendering so the residual forces and torques applied to the (virtual) object are used to compute an acceleration and hence to update the location of the object. Constraints such as a working surface like a table top are managed either as a force applied through a stiff element or as an impulse condition.

For most grasps friction forms an important and stabilising component of the force closure. Without friction the grasp would have to be naturally stable. This can be done by hanging a cup or mug by the handle with a hooked finger so that the centre of gravity can rest below the point of contact. In the absence of a convenient handle then at least 4 contact points are required to ensure that the object is restrained in three dimensional space. (Four points are a necessary condition given that the fingers can only apply a compressive force on the object). By including a consideration of the frictional forces of a grip, it is possible to manipulate the object with 5 degrees of freedom using only two points of contact. The degree of freedom that is not controlled is the rotation around an axis joining these two points of contact. A three finger grasp then becomes sufficient to constrain motion of the object within all 6 degrees of freedom.

2.2.1 Internal Models of Movement—Perceptions and Grasp

An important aspect of haptic rendering is the person's perception and how that can be used to promote realism. The concept of visual dominance is well documented [36, 37]. Giving visual and auditory cues can do much to add to the illusion of permanence in a haptic situation, for example, the authors noted that in the dental training workstation developed for the Haptel project (www.haptel.kcl.ac.uk) the variation of the pitch of the hand-piece as pressure was increased on the burr was a strong surrogate for the vibrations (that were missing) that would have been present in an operational drill.

These phenomena are explained by the fact that humans do a large degree of a-priory modelling about the nature of the object they are about to grasp, and in addition anticipate the loads during the manipulation [25]. Thus explaining perceptual phenomena such as

A-priory modelling

- Preshaping of grasp [20]
- Grasp illusions such as size-weight, texture-weight, grasp aperture-weight illusions (judgements of an objects weight are made well before movements start)

During manipulation

- Anticipation of object dynamics (grip force adjusts on lifting vs lowering an object) [17, 49]
- Correlation between grip force and load force. The grip force is modulated to be 'just enough' [23]

2.3 Friction Models for One Point Contacts

Friction is a complex and nonlinear phenomena with a corresponding plethora of approaches to modelling its effect. Most undergraduate engineering textbooks will only discuss linear (viscous) friction in any depth where the force due to the relative movement of the frictional surfaces is assumed to be proportional to the relative velocity. That is $f = Bv$ where B is the constant of proportionality. Although this model allows techniques such as Laplace transforms to be applied to the analysis, it is not a sufficiently accurate model of friction to describe grasp and support the haptic rendering of multi-finger systems requiring object manipulations.

The classic Coulomb friction model (shown in part in Fig. 2.5) provides a better representation of friction for haptic rendering. The model is characterised by a single value and is most commonly expressed as

$$F = F_c \, \mathrm{sgn}(v)$$

where the function sgn v computes the sign of the relative velocity. The frictional force F is assumed to oppose relative movement and the defining constant, F_c may, in practice, differ for positive and negative directions of sliding.

In certain situations it becomes necessary to consider the discontinuity of Coulomb friction around the origin. Thus for example, the frictional force is indeterminate at $v = 0$ lying somewhere between $-F_c$ and F_c. The LuGre friction model [2] includes microslip states to model microslip phenomena when there is an applied force that is less than the limit of static friction. The LuGre model is a relatively sophisticated model and a more intuitive realisation is the bristle model [9] which considers micro movements during periods of zero relative velocity, to be modelled as the interaction of two bristle surfaces. The bristle model reduces to a simpler stress model known as the Dahl model [19].

A well known method for computations involving coulomb friction is to generate a friction cone (Fig. 2.6) based on the limits of static friction and to consider the resultant force on an object that is subject to friction. The cone of friction provides a easy method to determine if the object is sliding by observing the resultant force with respect to the cone. If the resultant is contained within the cone then the object

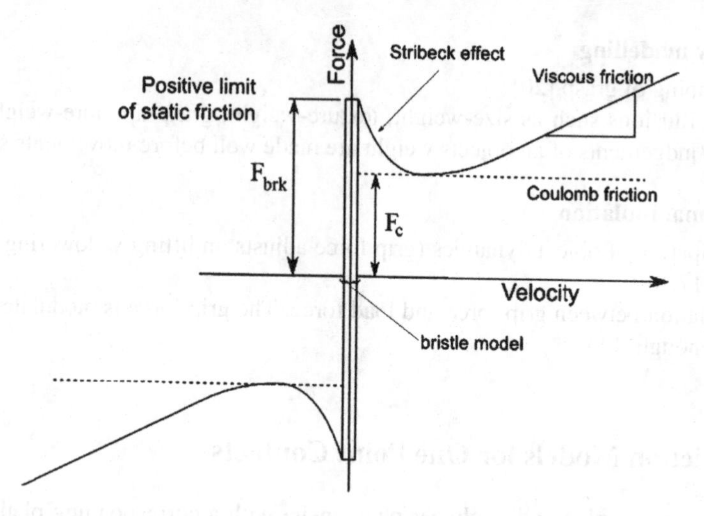

Fig. 2.5 Friction models shown as a function of velocity with a discontinuity at the origin to illustrate position phenomena

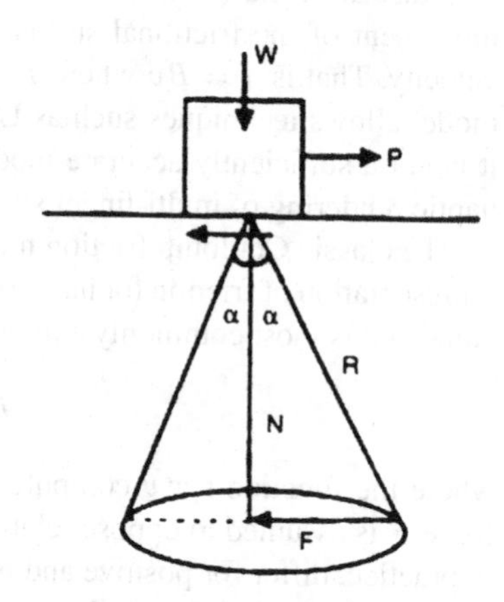

Fig. 2.6 Use of a cone of friction to determine if the object is sliding

is assumed to be at rest, if outside then it is assumed to be sliding. This concept is developed with respect to rendering of multi-finger contact systems in the following two sections.

2.4 Friction Cone Algorithm

The friction cone algorithm allows simple modelling of the friction between the fingers and the object to be built into the haptic rendering algorithm. It has the benefit of sufficient simplicity to be quick and efficient to implement, and sufficiently detailed to give the individual a good perception of manipulating common objects based on the friction needed to form simple multi-fingered grasps.

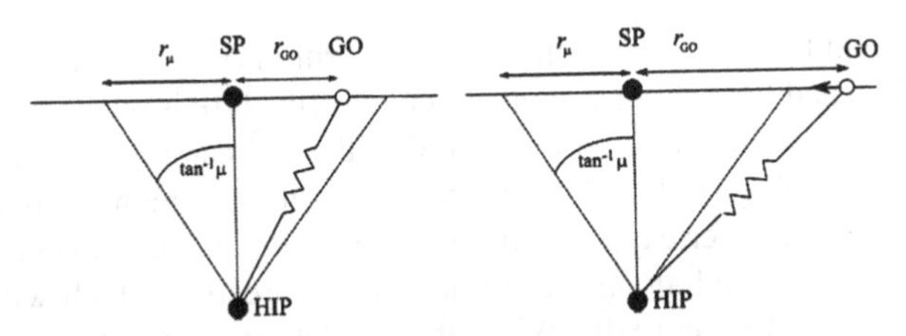

Fig. 2.7 Relationship between the god-object ($\overline{GO}$) and the bounding circle defined by the friction cone [31]. *Left*: shows the god-object within the circle and thus it remains static on the surface. *Right*: shows the god-object outside the bounding circle so it should be moved to the nearest point on the perimeter

The algorithm is based on the Salisbury and Zilles 'god-object' concept [51] which uses a point based god-object to maintain the history of contact between the person's fingers and the object.

The algorithm is described in detail in [31, 34] so is given in outline here.

A haptic interface point ($\overline{HIP}$) is defined as the representation of the end-point of the haptic interface in the virtual environment. (The overline convention is used where several letters are used to represent a vector variable such as $\overline{HIP}$.) In the case of finger contacts this is assumed to be one of the fingers involved in the manipulation of the object. A collision detection algorithm is needed to identify when the $\overline{HIP}$ is within an object, and at this stage it may be assumed that contact has been made. On first contact between the $\overline{HIP}$ and the object a notional god-object ($\overline{GO}$) is placed on the surface of the object and from thereon a force vector is assumed to be opposing the contact with a spring like characteristic. That is a force is created on the haptic interface that is proportional to the vector $\underline{v} = \overline{GO} - \overline{HIP}$. The position of the god-object is then used to model friction and this can be best described by considering the traditional friction cone with an apex half angle of $\tan^{-1}\mu$ but inverted so the apex is at the $\overline{HIP}$. This representation can be seen in Fig. 2.6. The god-object remains on the surface, but the additional constraint is now that it must also remain within the boundary of the circle formed by the intersection between the surface polygon and the friction cone.

As the $\overline{HIP}$ moves within the object the god-object may move outside the circle boundary. At this point the object is sliding over the fingers and this is handled within the algorithm by moving the god-object to the closest point on the circumference of the circle Fig. 2.7. A subsequent calculation may move the object with respect to the $\overline{HIP}$ and this would cause the surface point and the circle to move. Slip may then continue if this movement causes the god-object ($\overline{GO}$) to move outside the friction circle.

2.4.1 Extended Friction Cone Algorithm (xFCA)

Our initial work used a customised algorithm to identify collisions between the haptic interface point ($\overline{HIP}$) and the contacted object. The advantage of the god-object

($\overline{GO}$) method, and therefore the friction cone algorithm (FCA), is that once a collision is identified, and for as long as contact remains (the $\overline{HIP}$ is inside the object), no further collision detection passes are required as the $\overline{GO}$ can traverse the surface of the object. If the topology of the object is known (for the common triangle mesh this is a pre-computed edge/vertex connectivity graph) then on each update of the $\overline{HIP}$ position, the $\overline{GO}$ is moved along the surface in the direction which will minimise the distance between $\overline{GO}$ and $\overline{HIP}$. With efficient data structures the complexity of the mesh which can be traversed is bounded only by the available memory, and not by the processing speed. This method assumes that the initial collision is detected sufficiently quickly so that the $\overline{HIP}$ has not penetrated too deeply into the object. Often this calculation can occur at update rates significantly lower than typical for haptic rendering (0.8–1 KHz). See Melder and Harwin [32–34] and Melder [30] for a description of such a surface traversal algorithm.

However, it is often desirable that the $\overline{HIP}$ is not a single point and in fact has some volume. This reduces the problems of the point escaping through cracks between poorly generated surface triangles and facilitates the use of convex-convex collision detection libraries [11, 38, 46]. In this case surface traversal cannot be easily computed. If the two colliding objects can be approximated as convex hulls, then fast and predictable solutions for surface traversal can be implemented. One such method is to compute the Minkowski Sum of the two convex objects and proceed to traverse this combined surface as a single point [38], however, computing the Minkowski Sum becomes cumbersome for non trivial shapes undergoing rotations.

As efficient algorithms for collision detection between convex polyhedral are now widely available though well supported software libraries, such as SOLID [47] (dtecta.com) and Bullet [11] (bulletphysics.org) we prefer a more generalisable method of rendering convex-convex haptic interaction which we refer to as the extended friction cone algorithm xFCA. While the convenience of not having to perform repeated collision passes during haptic rendering is lost, convex-convex collision detection is efficient even for very high polygon count models if the convex hull is computed [46].

The xFCA algorithm can be summarised as follows:

1. When the $\overline{HIP}$ is in free space, the haptic cursor is mapped to that position, Fig. 2.8-1. At each simulation update the haptic cursor is moved to the new $\overline{HIP}$ position and a full collision pass is performed looking for any contact between the haptic cursor and other objects.
2. If a contact is found, Fig. 2.8-2, the god-object is created for that $\overline{HIP}$-Object pair and the main xFCA loop begins. The haptic cursor remains with the $\overline{GO}$ on the surface.
3. During the same simulation frame, the penetration depth and normal returned from the collision detection system is used to move the $\overline{GO}$ to the surface of the object, Fig. 2.8-3. The new surface position of the $\overline{GO}$ is stored, along with the surface normal and plane at the point of contact.
4. The simulation now proceeds to the next frame. If the $\overline{HIP}$ has moved with respect to the previous frame, the position of the GO must be updated to reflect the new surface position. Before performing a collision pass, the $\overline{GO}$ is moved

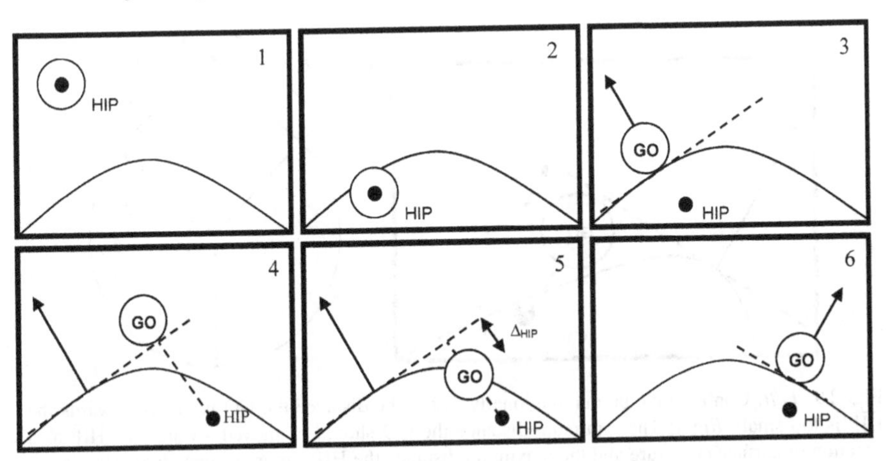

Fig. 2.8 The progression of the xFCA moving the god-object (GO) along a surface and the haptic interaction point ($\overline{\text{HIP}}$) into the object

along the surface plane to the point at which it is closest to the $\overline{\text{HIP}}$, Fig. 2.8-4, by removing that portion of $\underline{v}$ (the $\overline{\text{GO}}$ to $\overline{\text{HIP}}$ vector) which is in the direction of the surface normal $\mathbf{n}_s$:

$$\overline{P}^{+}_{GO} = \overline{P}_{GO} - (\underline{v} \cdot \mathbf{n}_s)\mathbf{n}_s$$

where $\underline{v} = \overline{\text{HIP}} - \overline{\text{GO}}$.

5. Next, the $\overline{\text{GO}}$ is moved down towards the $\overline{\text{HIP}}$ by a fixed amount, Δ_{HIP}, in order that the $\overline{\text{GO}}$ should be beneath the object's surface so the collision pass will return a positive contact, Fig. 2.8-5. There is more to this step which will be discussed next.
6. A collision pass is now performed to return a new penetration depth and contact normal for the GO-Object collision which is then used to reposition the $\overline{\text{GO}}$ on the surface of the object, Fig. 2.8-6.

Stages 3–6 continue until either, the $\overline{\text{HIP}}$ crosses the surface plane, or the collision pass fails to find contact between the $\overline{\text{GO}}$ and the object. Both of these conditions are assumed to indicate that the $\overline{\text{HIP}}$ has lost contact with the object. At the end of each xFCA cycle, once the $\overline{\text{GO}}$ is positioned on the true object surface, the distance between the $\overline{\text{GO}}$ and the $\overline{\text{HIP}}$ is used to calculate a spring force which is applied to the haptic device and the virtual object.

The reason the $\overline{\text{GO}}$ is not moved the full distance to the $\overline{\text{HIP}}$ is that for thin objects or large forces this would result in push through. However, if it is too small compared to the radius of curvature of the surface then the $\overline{\text{GO}}$ may fail to retain contact even though the $\overline{\text{HIP}}$ is still within the object, Fig. 2.9, left.

If the maximum distance that the $\overline{\text{HIP}}$ will move in a single time step, d_{max}, is known, it is possible to approximate the minimum Δ_{HIP} necessary to guarantee sustained contact on a surface with radius of curvature r_c, Fig. 2.9, right. This results

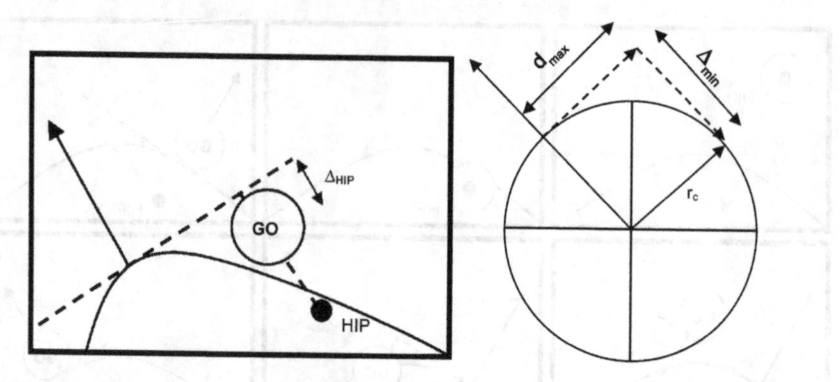

Fig. 2.9 *Left*: Contact may be unintentionally lost if the distance the $\overline{GO}$ is moved towards the $\overline{HIP}$ is too small. *Right*: The minimum distance the $\overline{GO}$ should be moved towards the $\overline{HIP}$ as a function of surface curvature and the maximum distance the $\overline{HIP}$ can move in a single simulation step

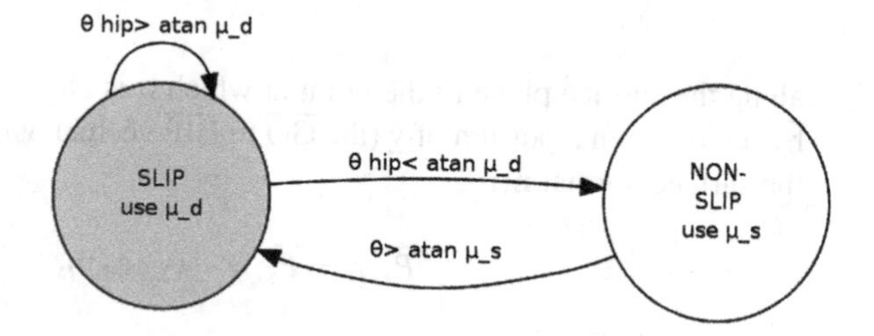

Fig. 2.10 Finite state machine to track transitions between slipping (dynamic coefficient of friction) and static conditions

in the following relationship:

$$\Delta_{HIP} = \sqrt{2r_c^2 - (d_{max})^2}$$

The value of Δ_{HIP} can also be used, along with the smallest dimension of the haptic cursor, to define the minimum depth an object can be penetrated before push-through becomes possible. In practice, the common haptic update rate of 1000 KHz or more results in very small values of d_{max}, even at high velocities.

2.4.1.1 Modelling Dynamic and Static Friction

Implementing static friction requires a slight modification from the original FCA. The main difference is that, if the friction cone is large compared to d_{max} when it is exceeded, the requirement for a bounded dmax is broken and subsequently the $\overline{GO}$ may be moved a large distance away from the surface causing a loss of contact. To solve this we introduce the concept of the Surface Object (SO). The SO represents the theoretical position of the $\overline{GO}$ if there were no friction, is updated at every time step and is assumed not to lose contact unless the $\overline{HIP}$ leaves the surface. However, as it is the GO which represents the true current surface position, contact is only lost

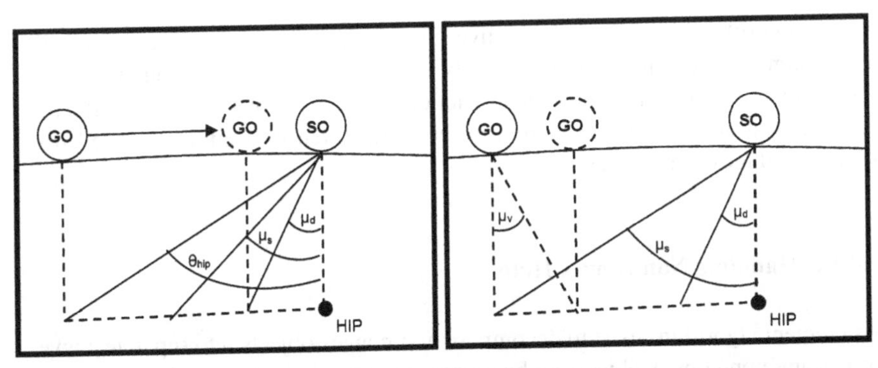

Fig. 2.11 *Left*: If using Coulomb friction, when the $\overline{GO}$ exits the static friction cone it is placed on the edge of the, smaller, Coulomb friction cone. *Right*: To create a force proportional to speed (viscous friction) the $\overline{GO}$ is moved a maximum distance proportional to penetration depth at each time step bounded by the Coulomb friction radius

if the $\overline{HIP}$ crosses the GO's plane and not that associated with the SO. Should the $\overline{SO}$ lose contact at any time, it remains positioned at the last known point of contact on the surface until either another surface position is found or the $\overline{HIP}$ transitions the GO's contact plane.

An easy adaption to this algorithm allows a simplified version of Stribeck's effect to be implemented. If the finger is not sliding with respect to the object the algorithm is computed based on a static coefficient of friction, but if slipping is occurring then a dynamic coefficient of friction is used. The finite state machine shown in Fig. 2.10 is used to track the friction state and considers only the two conditions, slipping or static.

When the angle θ_{HIP} exceeds the static friction angle the GO is moved to the edge of the Coulomb friction cone defined by μ_d where $\mu_d < \mu_s$, Fig. 2.11, left. At each subsequent time step, if the $\overline{GO}$ is outside the Coulomb friction cone it is moved back to the nearest edge. If the $\overline{GO}$ is found inside the Coulomb friction cone, it is left in place. This gives a stick-slip effect with a force proportional to normal force during surface slip.

Viscous friction can also be neatly integrated into the same framework. Viscous friction is proportional to the surface velocity as well as normal force and requires that the $\overline{GO}$ is moved a fixed distance relative to its own position, not the $\overline{SO}$ position. From the friction model in Fig. 2.7 it is clear that Coulomb friction defines a base level of friction during surface slip and viscous friction increases it above this level relative to velocity. To mimic this, during the slip phase of the friction model, the $\overline{GO}$ is moved towards the edge of the Coulomb friction circle but only as far as the edge of the viscous friction circle centred on the GO. This means that, at each time step, if the $\overline{SO}$ is moved away from the $\overline{GO}$ by less than the radius of viscous friction only Coulomb friction will be felt. If the surface velocity increases, the position of the $\overline{GO}$ will begin to trail behind creating a greater frictional force, Fig. 2.11, right.

The algorithm is simple and effective. Simple because changing between static and dynamic coefficients of friction simply alters the diameter of the bounding circle formed by the intersection between the friction cone and the object surface. Effective because it allows the person to reduce their grip force to a minimum needed to support and move the object.

2.4.1.2 Handling Non-convex Hulls

As efficient algorithms exist to decompose non-convex objects into separate convex pieces any non-convex object can be rendered haptically in this manner by treating it as multiple separate HIP-Object contacts, each with its own local $\overline{GO}$ stored to calculate forces arising due to penetration depth. However, it should be noted that it becomes possible to slide into gaps between convex pieces due to the force from opposing $\overline{GO}$s cancelling each other out, though this can be minimised with careful convex-decomposition.

Certain classes of object do not decompose easily into large convex pieces, smooth concave curves for example (coffee mug handle). If rendering quality is to be preserved in such cases, or discontinuities such as slipping into cracks between convex pieces are to be removed entirely it is necessary to perform rendering directly between non-convex objects. For the xFCA this would mean computing the penetration depth of intersecting non-convex polyhedra, a non-trivial task which can be, in the worst case, $O(nm)$ for the number of polygons in each mesh [22].

It is possible to make use of the fact that distance calculation between non-overlapping non-convex objects is achievable in real time and hence some researchers employ a haptic rendering algorithm based on repulsion rather than penetration to avoid the situation of overlapping non-convex objects [24]. Without very high stiffness, this technique results in the haptic cursor not visibly contacting the object which may not be desirable in realistic simulation.

The technique used by the authors for general non-convex haptic rendering is similar to this repulsive technique but is in fact closer to the constraint based methods used in animation and games physics engines and in fact may be considered an intermediary step between simply coupling haptic interfaces directly into constraint based physics engines and removing a separate haptic rendering stage entirely. Although this is straightforward algorithmically it is not yet widely employed because, at the time of writing, computing power is not yet sufficient to allow common game physics engines, such as the Bullet library, to run at real time speeds with haptic time steps (<1 mS) even for a seemingly simple virtual environment.

The principle of the technique is now briefly described. Using this method, the haptic cursor is treated as a very light mass object coupled to the $\overline{HIP}$ by a spring with a high stiffness and a damper and its position is updated each time step based on standard Newton dynamics equations as described in Sect. 2.5.1. As the mass is so light, and the spring stiffness comparatively high, when in free space no forces are perceived by the user as the haptic cursor does not lag far behind the $\overline{HIP}$ (some tuning of the damping constant is required to prevent vibrations). The haptic cursor is continually checked against other objects in the environment for a collision. Should a collision occur, a frictionless constraint based linear complimentary problem (LCP) model is formed and solved to calculate the forces and torques which, when applied to the haptic cursor, will exactly cancel out acceleration in the direction of further penetration and effectively cause the objects to slide over each other. Formulation of this type of LCP for object dynamics is a well studied problem, see the original work by Baraff [3] for a simple explanation and formulation, though many more efficient implementations have since been proposed [1, 43].

Depending on the requirements of the simulation, the force based interaction between the haptic cursor and contacted object may be entirely decoupled from the wider physics engine which governs movement of dynamic objects in the scene. From the point of view of the haptic rendering constraint model it is operating on an object which is fixed in space. From the point of view of the physics engine, the haptic cursor does not exist except as a disturbance force added into the global force matrix. The result is that even if the physics engine slows temporarily, the haptic rendering proceeds uninterrupted and the perception of the user is of the inertia of the contacted object(s) increasing.

Implementing a combined penalty and constraint based haptic rendering algorithm in this manner is practical because solving the LCP resulting from contact between only two objects where one is stationary is achievable within the haptic time steps even when many points of intersection are found.

However the following points should be taken into consideration:

- Calculation of friction between the contacted object and haptic cursor is approximated using the FCA as the addition of the extra constraints into the LCP contact problem results in unpredictable solution times.
- It is believed that newer techniques for approximating friction in LCP based object dynamics may solve this but it is untested by the authors.
- To maintain haptic update rates it is preferred that the haptic cursor should be decomposable into convex parts or the collision detection between complex non-convex objects will be too slow for haptic rendering

2.5 Haptic Rendering and Manipulation of Virtual Objects

Multi-finger contact in a virtual environment allows complex and direct manipulation of physical objects. Any haptic rendering algorithm requires a two stage physics engine. Stage 1 is to identify the forces within the system of objects due to any points of contact or collisions, stage 2 is to apply the appropriate physical laws to compute the response of these objects. Stability of the haptic device requires rapid computation of the response of the simulated physics to the applied forces to those objects with a direct attachment to the person (through the haptic device). The following discussion will be restricted to rigid object manipulation where the object shape, mass and inertia remains unchanged through the manipulation. Extending the work

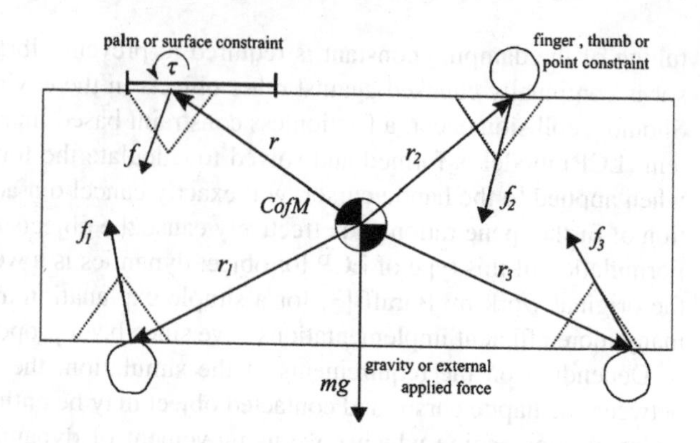

Fig. 2.12 Residual forces and torques on a grasped object

to rendering flexible objects with visco-elastic properties and the ability to reshape the object through cutting, wearing and drilling during the rendering remains a difficult challenge. However for the most part this application area is sufficiently well covered by haptic interfaces using a single point of contact via a surrogate tool, so has thus far not demanded academic attention.

The friction cone algorithm described naturally provides a key piece of information for multi-finger manipulation of rigid physical objects, that is the force vector at each contact point, that is to say each finger. This force vector is directed from the HIP to the GO and is proportional to the length of this vector. The proportional constant is directly related to the perceived stiffness of the surface of the rigid object. Manipulation of the object now becomes, in essence, the application of Newton-Euler dynamics to the virtual object.

Consideration of the manipulation of the object is most conveniently done by separating translation and rotation of the object, and applying Newton's law to the former and Newton-Eulers law to the latter case. If the object is at rest the sum of all force vectors on the object must be zero. Should any of the forces applied to the object deviate, this will give rise to a small unbalanced residual force which, once determined, can be used to calculate the acceleration of the object and hence by integration, the updated velocity and position. Forces on the object can arise from finger contacts, contact between the object and other objects, or forces due to gravity. The algorithm will of course still work if the author chooses to represent gravity as an upward acceleration of the object, but the normal convention is to consider gravity as a constant force vector with the direction fixed in the universal frame.

2.5.1 Computation of Translation from the Residual Forces

Considering the forces shown in Fig. 2.12 we can sum the forces on the virtual object. To implement Newton's law we must give the object a notional (real) mass m, and this mass can be considered to be entirely located at the object centre of gravity which can be located within a coordinate frame attached to the object. It is usually convenient to make the centre of mass the origin of this coordinate frame for both the translational and rotational calculations. From Fig. 2.12 it can be seen that the residual force is calculated as the sum of forces on the object, in this illustration this is

$$\underline{F}_{\text{residual}} = \sum_{i=1}^{n} \underline{f}_i + \underline{g}$$

Newton's equation is now used to compute the acceleration so

$$\underline{a} = \underline{F}_{\text{residual}}/m$$

and from the acceleration the velocity and position are readily calculated by Euler integration as

$$\underline{v}_{new} = \underline{v}_{old} + \underline{a}\Delta_t \tag{2.1}$$

$$\underline{p}_{new} = \underline{p}_{old} + \underline{v}_{new}\Delta_t \tag{2.2}$$

Δ_t is the loop time of the control algorithm. In general Δ_t needs to be small since large computational delays result in haptic instabilities [10]. In practice most hardware dictates that Δ_t is of the order 1 to 2 milliseconds.

2.5.2 *Computation of Rotations from the Residual Torques*

A similar but slightly more complex approach is used to compute and update the rotation parameters for the object coordinate frame. A variety of methods are available to specify the rotation of the object coordinate frame with respect to the base frame. A common method in computer graphics and robotics is to specify rotations as a 3×3 orthogonal matrix that can be used on a 3×1 vector or n vectors expressed as a $3 \times n$ matrix of vectors. This rotation matrix will rotate the components of vectors expressed in the object's coordinate frame into in components in the base frame. This rotation matrix has a dual use, the first to express this coordinate frame rotation, the second (and related use) is to convert the objects inertial matrix J from the local to the base frame.

The full Newton-Euler relationship is

$$\underline{T} = J\underline{\dot{\omega}} + (\underline{\omega} \times J\underline{\omega})$$

where T is the opposed torque, $\times$ represents the cross product, and J the object inertia matrix. However the second term in this equation $(\underline{\omega} \times J\underline{\omega})$ contributes the 'gyroscopic term' and in practice can be ignored. Using this simplified Newton-Euler relationship we can rearrange the equation to compute the angular acceleration from the object inertia and the computed torque. That is

$$\underline{\dot{\omega}} = \underline{T}J^{-1} \tag{2.3}$$

A similar technique to the calculation of the residual forces can be used to compute the residual torque on the object, that is—to use the vector between the haptic interface point ($\overline{\text{HIP}}$) and the god-object ($\overline{\text{GO}}$) to represent the forces applied to by the fingers. These forces represent a set of torques, conventionally computed around the centre of gravity, that can be summed to estimate the residual torque causing the angular acceleration, and hence the angular velocity and ultimately the rotational terms expressed in the matrix R.

Other sources of torque such as area of contacts, can be added to the torques due to the finger contacts, before the computation of angular acceleration is made, thus the residual torque ($\underline{T}_{\text{residual}}$) is

$$\underline{T}_{\text{residual}} = \sum_i r_i \times f_i + \sum_i \tau_i$$

where $r \times f$ are the torques due to individual point contacts, usually assumed to be point representations of the finger contacts, and τ are the torques applied through area contacts.

An inertia matrix is needed to complete the calculation in (2.3). This is most conveniently retained as the inertia along the three principal axes of the object, and these are often aligned with the local object coordinate frame. The inertia is stored in a matrix form that is

$$A = \begin{bmatrix} I_{xx} & 0 & 0 \\ 0 & I_{yy} & 0 \\ 0 & 0 & I_{zz} \end{bmatrix}$$

A prior estimate of the rotation matrix is needed to convert this inertia matrix into the matrix representation the instantaneous rotation, that is

$$^{A}J = RAR^{T}$$

The full calculation for the angular acceleration is thus

$$\dot{\underline{\omega}} =^{A} J^{-1}\left(T - \underline{\omega} \times^{A} J\underline{\omega}\right)$$

but is simplified to ignore the ‘gyroscopic term’ to be

$$\dot{\underline{\omega}} \approx RJ_p^{-1}R^{T}T$$

$\dot{\underline{\omega}}$ represents the angular acceleration around the three axes of the base coordinate frame, and these can be treated as independent for the first integration to compute the instantaneous angular velocity. That is

$$\underline{\omega}_{new} = \underline{\omega} + \dot{\underline{\omega}}\Delta_t$$

However the second integration needs to compute the absolute position of the object, and to this end is calculated in matrix form. The first stage is to compute the skew matrix S

$$S = \begin{bmatrix} 0 & -\omega_z & \omega_y \\ \omega_z & 0 & -\omega_x \\ -\omega_y & \omega_x & 0 \end{bmatrix}$$

S is a convenient method to represent instantaneous angular velocities since it can be shown that $\dot{R} = SR$.

The derivation of this expression follows by differentiating the definition of orthogonal matrix $RR^T = I$ and a good discussion of this is given in Craig [13]. If integration happens over a sufficiently small interval then the rotation matrix can be written as

$$\dot{R} \approx \Delta R = I + S\Delta_t$$

The Euler integration is then

$$R_{new} = R\Delta R = R(I + S\Delta t)$$

where the new rotation matrix is R_{new} and the new vector of rotational speeds is $\underline{\omega}_{new}$.

2.6 Multi-body Dynamics for Haptic Rendering

Realistic simulation of multiple moving bodies for natural object manipulation with haptics is computationally challenging. A full discussion of implementation issues in multi-body dynamics simulation is out of the scope of this chapter, although the following discussion is based on Baraff [3], Anitescu and Potra [1] and Stewart and Trinkle [43]. The following points should be considered when implementing multi-body physics simulation for haptics:

- At haptic update rates, even the most efficient algorithms are unlikely to be fast enough to maintain real-time solutions for large numbers of objects
- The haptic sense is highly sensitive to force discontinuities
- The common shortcuts used in games physics engines are likely to be manifested as force discontinuity and as such will be detected and perceived by the user as an inconsistency
- The haptic sense is sensitive to all objects in physical contact, thus the haptic rendering must consider the energy coupling through all objects connected to the physical haptic device. An example would be the haptic rendering of a tray carrying a tea service.

Considering these points, a framework for a multi-body haptics simulation can be formulated to maximise the use of computing resources and minimise discontinuities perceptible to the user.

The current computing trend towards parallel computing architectures means that highly multi-threaded software solutions can be effectively employed to help

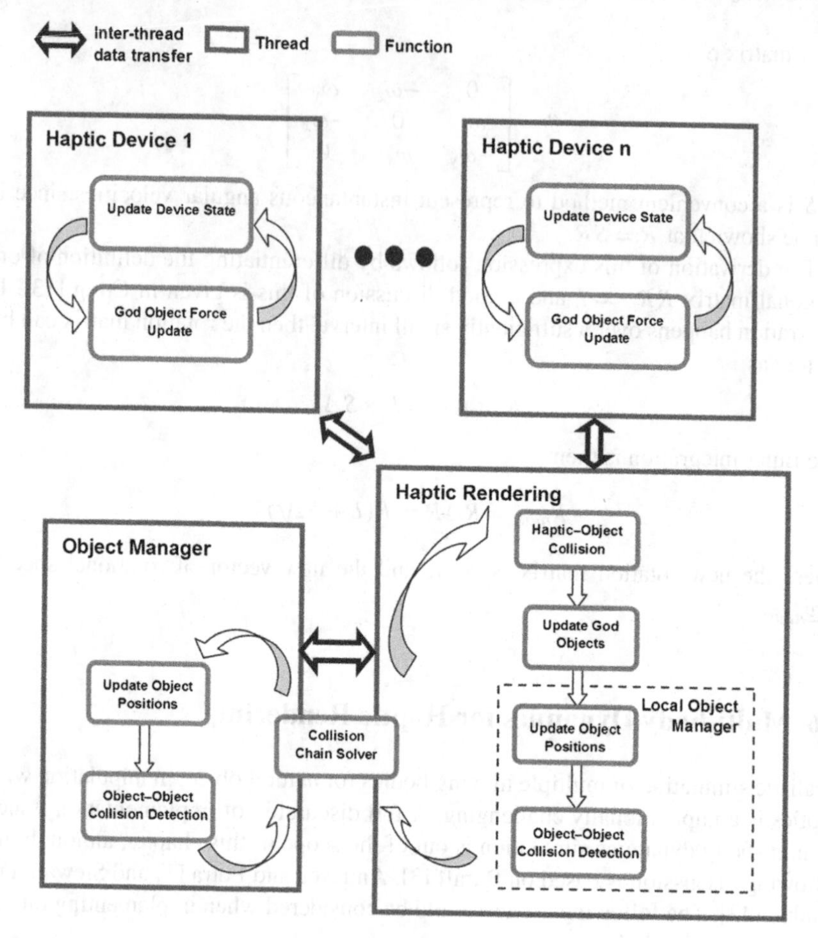

Fig. 2.13 Example threading model and information interchange for a multi-body multi-finger haptic simulator

achieve the real-time needs of a haptically enabled simulation. It is common to separate graphics rendering into its own or multiple threads and haptics and physics rendering can be treated in a similar manner. An example multi-threading structure for a multi-body haptic simulator is shown in Fig. 2.13.

2.6.1 Multi-threaded Haptic Renderer

In this model, a strict threading priority hierarchy is maintained and an important aspect of the implementation is that during interchange of thread safe information, no thread is ever made to wait on a lower priority thread for a non-deterministic length of time, usually the time required to make a copy of a thread safe data structure.

Each haptic device has its own dedicated thread, the highest priority, which communicates with the device, updates the force commanded based on the current stored location of any god-objects associated with it but does not update the god-objects themselves. This means that should the haptic rendering ever slow, stability of the haptic device is not compromised, but an increase in dynamic friction may be perceived.

The next highest priority threads are those associated with haptic rendering, i.e. collision detection between haptic cursors and scene objects and the movement of the god-object along the surface of a contacted object using the FCA. The haptic rendering threads may also be tasked with updating a small number of contacted objects, the reasons for this are discussed next.

2.6.2 Multi-threaded Object Manager

The object manager threads are responsible for the motion of dynamic objects. There are three parts to this: the integrator, which calculates future velocities and positions of objects based on the accumulated forces and torques acting on them, the collision detection system, which reports the points of contact, and the constraint handler, which calculates appropriate forces, torques and impulses so, on the next integrator pass, the motion of objects does not result in further interpenetration.

The rigid-body dynamics algorithm has the following steps:

1. Apply all external forces acting on objects (haptics, gravity etc.)
2. Process all objects and find all points of contact
3. Group all contacts into collision chains
4. Solve each collision chain to find new accelerations and velocities
5. Update each object's state vector and repeat

External forces are applied first so that they are taken into account when the contact matrices are formulated and solved. Once all the points of contact have been identified they are grouped into collision chains. Contact points are part of the same collision chain if they are connected via a movable object. Two movable objects sitting apart but on the same static floor would be part of separate collision chains. If the movable objects were touching they would be part of the same collision chain.

The purpose of separating objects into collision chains is that each chain is an independent system and can be solved in parallel. Once all contact points are grouped into chains they are assigned to separate threads to be solved. A pool of pre-existing threads is used to remove the overhead associated with creating and destroying them.

As the object dynamics is decoupled from the haptic rendering it can be updated at a different rate. Ideally, the system will attempt to maintain as high a frame rate for the physics as possible by using a variable length time step. However, if the contact solver begins to slow too much, the step size will grow large enough for the user to feel objects judder instead of move smoothly, this is likely to happen long before it can be detected visually. To prevent this, the time step is capped at 0.002 ms (500 Hz). If the previous time step took longer, the simulation will fall behind real-time rather than take larger and larger steps. The advantage of this method is that neither the objects nor the haptic rendering will become unstable, if the object update slows significantly then the user will perceive an increase in the apparent inertia of the object but will still be free to haptically explore the object.

Once a haptic device makes contact with an object, if required the rendering thread can take over responsibility for updating it. Part of the haptic rendering loop now includes the same functions as the object manager and any objects assigned to run at the haptic update rate are processed here. If the object being updated by the haptic thread makes contact with another object, it too becomes the responsibility of the haptic thread. When the number of objects in the collision chain assigned to the haptic thread exceeds a predefined safe limit (usually very low, 1 to 3 objects) for guaranteeing update rate, the whole chain reverts to the responsibility of the Object Manager.

The further advantage of this technique is that stability problems associated with grasps involving objects with very low rotational inertia can be minimised by updating their position at haptic update rates. If multiple objects are in contact so that the chain cannot be updated at haptic rates, the increased inertia and energy loss due to the stack of objects are still likely be sufficient to remove and instabilities.

2.7 Discussion

The evidence for providing good multi-finger multi-contact haptic interactions is clear, although many technical challenges remain. The transparency of the haptic device is important, and the approach of using low inertia haptic interfaces helps to achieve this. However this is at the expense of workspace so it has proved difficult to produce transparency in a large working volume. The approach by Barrow and Harwin [5] partially addresses the issue of transparency, workspace and high achievable stiffness but further development along this line would introduce higher costs and greater need to ensure the operator's safety. Smaller multi-finger workspaces are more readily achievable, in particular with the potential fall in cost of high quality devices such as the W3D and W5D produced by Entact Robotics (entactrobotics.com), the Novint Falcon and the Sensible Omni, although all three would require the tool surrogate to be replaced by a more appropriate interaction point such as a thimble. A strong limitation is self interference of the multiple haptic interface and although some work has been done in this area, more work is needed to allow rapid evaluations of the trade-off between the movements required by a range of tasks, and the capabilities of the device. For example most kinematics would not allow the operator to 'cross fingers'. In our own work on stroke rehabilitation, the natural workspace expected by the therapists included a full range of movement that included shoulder rotation. No device has yet achieved this capability, while still allowing individual free movement of the individual fingers and thumbs.

The algorithm discussed in this chapter provides a good method for doing multi-contact multi-finger interactions that is independent of device kinematics and will

Fig. 2.14 Illustration of multi-point contacts using two Phantom 1.5 haptic interfaces

work across a wide range of haptic devices. The threading approach ensures that high update speeds can occur on those parts of the system that require rapid information for stability, while still maintaining good visual rendering and object dynamics for the objects that are not contacted by the haptic device and operator. The method is robust to contact instability (due to threading) in the sense that dynamics of objects and haptic rendering are not decoupled but not reliant on one another. Should the object update slow it is simply perceived as an increase in the inertia. The method requires no models, and is such that objects held in a grasp with any number of contact points tend to centre in the grasp via the dynamics of the object. Thus transportation of the objects appears natural. The experimental setup used in this work is shown in Fig. 2.14 and there are also videos of the device published on Youtube under the user ReadingTHRIL.

As the technology moves forward these algorithms must be adapted to multi-contact interactions where there is significant delay, such as Internet delayed signal paths, as well as allow more complex material interactions such as soft body physics including the ability to cut, tear, drill and probe simulated materials such as tendons, muscles, bone, teeth, clays, woods, metals etc.

2.8 Conclusions

There are still many problems to be addressed in providing good multi-finger multi-contact haptic interactions. These include cost, workspace, transparency and device stiffness. However if these can be addressed there are likely to be a wide range of applications, ranging from education through to data interaction and computer interfacing. There is a need to understand the mechanisms and neuroscience behind grasp to ensure that the device is transparent to the tasks the operator is trying to perform and thus a brief overview of this was presented. As computing advances so must the computer interfaces and metaphors for user interactions must change. In the words of Gauldie, Wright and Shillito, *3D modelling is not for wimps* [18, 39]. The algorithms discussed in this chapter allow for interactions to occur in 3D. Indeed the method allows a world where all digits of both hands can be used to manipulate and interact with rigid objects.

Acknowledgements Many individuals have contributed to this work, including technical and academic staff, as well as members of the tHRIL laboratory. The authors are pleased to acknowledge in particular the contributions made by Dr Nic Melder and Mr Sebastian McKnight.

References

1. Anitescu, M., & Potra, F. A. (1997). Formulating dynamic multi-rigid-body contact problems with friction as solvable linear complementarity problems. *Nonlinear Dynamics*, *14*(3), 231–247.
2. Åström, K. J., & de Wit, C. C. (2008). Revisiting the LuGre friction model: stick-slip motion and rate dependence *IEEE Control Systems Magazine*, *28*(6), 101–114.
3. Baraff, D. (1994). Fast contact force computation for nonpenetrating rigid bodies. In *Proceedings of the 21st annual conference on computer graphics and interactive techniques* (pp. 23–34). New York: ACM.
4. Barbagli, F., Salisbury, K. Jr, & Devengenzo, R. (2003). Enabling multi-finger, multi-hand virtualized grasping. In *Proceedings of the 2003 IEEE international conference on robotics and automation, ICRA'03* (Vol. 1, pp. 809–815). New York: IEEE Press.
5. Barrow, A., & Harwin, W. (2008). High bandwidth, large workspace haptic interaction: flying phantoms. In *Haptic interfaces for virtual environments and teleoperator systems*.
6. Brooks, F. P. Jr. (1999). What's real about virtual reality? *IEEE Computer Graphics and Applications*, *19*(6), 16–27.
7. Brooks, F. P. Jr., Ouh-Young, M., Batter, J. J., & Kilpatrick, P. J. (1990). Project GROPEHaptic displays for scientific visualization. In *ACM SIGGraph computer graphics* (Vol. 24, pp. 177–185). New York: ACM.
8. Bullock, I. M., & Dollar, A. M. (2011). Classifying human manipulation behavior. In *IEEE international conference on rehabilitation robotics (ICORR 2011)* (pp. 532–537). New York: IEEE Press.
9. Canudas de Wit, C., Olsson, H., Astrom, K. J., & Lischinsky, P. (1995). A new model for control of systems with friction. *IEEE Transactions on Automatic Control*, *40*(3), 419–425.
10. Colgate, J. E., & Schenkel, G. G. (1994). Passivity of a class of sampled-data systems: application to haptic interfaces. In *American control conference* (pp. 3236–3240).
11. Coumans, E., et al. (2006). Bullet physics library. Open source: bulletphysics.org.
12. Cox, M. J., Quinn, B. F., Newton, J. T., Banerjee, A., & Woolford, M. (2012). Researching haptics in higher education: the complexity of developing haptics virtual learning systems and evaluating its impact on students' learning. *Computers & Education*, *59*(1), 156–166. doi:10.1016/j.compedu.2011.11.009.
13. Craig, J. J. (1989). *Introduction to robotics: mechanics and control*. Reading: Addison-Wesley. ISBN 0-201-09528-9. UR call 629.892-CRA.
14. Cutkosky, M. R. (1989). On grasp choice, grasp models, and the design of hands for manufacturing tasks. *IEEE Transactions on Robotics and Automation*, *5*(3), 269–279.
15. Endo, T., Kawasaki, H., Mouri, T., Ishigure, Y., Shimomura, H., Matsumura, M., & Koketsu, K. (2011). Five-fingered haptic interface robot: HIRO III. *IEEE Transactions on Haptics*, *4*(1), 14–27.
16. Flanagan, J. R., & Wing, A. M. (1990). The stability of precision grip forces during cyclic arm movements with a hand-held load. *Experimental Brain Research*, *105*(3), 455–464.

17. Flanagan, J. R., & Wing, A. M. (1997). The role of internal models in motion planning and control: evidence from grip force adjustments during movements of hand-held loads. *The Journal of Neuroscience*, *17*(4), 1519.
18. Gauldie, D., Wright, M., & Shillito, A. M. (2004). 3D modelling is not for WIMPS part ii: stylus/mouse clicks. In *Proceedings of eurohaptics* (pp. 182–189).
19. Haessig, D. A. Jr, & Friedland, B. (1991). On the modeling and simulation of friction. *Journal of Dynamic Systems, Measurement, and Control*, *113*(3), 354–362.
20. Haggard, P., & Wing, A. (1995). Coordinated responses following mechanical perturbation of the arm during prehension. *Experimental Brain Research*, *102*, 483–494.
21. Hyun, J.-W., Kumazawa, I., & Sato, M. (2007). A new measurement methodology of multi-finger tracking for handheld device control using mixed reality. In *Annual conference, SICE 2007* (pp. 2315–2320). New York: IEEE Press. doi:10.1109/SICE.2007.4421375.
22. Jiménez, P., Thomas, F., & Torras, C. (2001). 3d collision detection: a survey. *Computers & Graphics*, *25*(2), 269–285.
23. Johansson, R. S., & Westling, G. (1984). Roles of glabrous skin receptors and sensorimotor memory in automatic control of precision grip when lifting rougher or more slippery objects. *Experimental Brain Research*, *56*(3), 550–564.
24. Johnson, D. E., & Willemsen, P. (2003). Six degree-of-freedom haptic rendering of complex polygonal models. In *Proceedings of the 11th symposium on haptic interfaces for virtual environment and teleoperator systems, HAPTICS 2003* (pp. 229–235). New York: IEEE Press.
25. Körding, K. P., & Wolpert, D. M. (2004). Bayesian integration in sensorimotor learning. *Nature*, *427*, 244–247.
26. Light, C. M., Chappell, P. H., & Kyberd, P. J. (2002). Establishing a standardized clinical assessment tool of pathologic and prosthetic hand function: normative data, reliability, and validity. *Archives of Physical Medicine and Rehabilitation*, *83*, 776–783.
27. Loureiro, R. C. V., & Harwin, W. S. (2007). Reach and grasp therapy: design and control of a 9-dof robotic neuro-rehabilitation system. In *Proceedings of the 2007 IEEE 10th international conference on rehabilitation robotics*, 12–15 June 2007 (pp. 757–763).
28. Loureiro, R. C. V., Lamperd, B., Collin, C., & Harwin, W. S. (2009). Reach and grasp therapy: effects of the gentle/g system assessing sub-acute stroke whole-arm rehabilitation. In *IEEE international conference on rehabilitation robotics* (pp. 755–760).
29. McKnight, S., Melder, N., Barrow, A. L., Harwin, W. S., & Wann, J. P. (2005). Perceptual cues for orientation in a two finger haptic grasp task. In *First joint eurohaptics conference and symposium on haptic interfaces for virtual environment and teleoperator systems*, 18–20 March (pp. 549–550). New York: IEEE Press.
30. Melder, N. (2011). *Multi-finger manipulation physics for haptic rendering*. PhD thesis, University of Reading. Call number, THESIS-R10821.
31. Melder, N., & Harwin, W. (2002). Improved rendering for multi-finger manipulation using friction cone based god-objects. In S. Wall, M. Wright, & A. M. Shillito (Eds.), *Proceedings of eurohaptics conference* (pp. 82–85). Edinburgh College of Art and University of Edinburgh.
32. Melder, N., & Harwin, W. S. (2004). Extending the friction cone algorithm for arbitrary polygon based haptic objects. In *Haptic interfaces for virtual environment and teleoperator systems* (pp. 234–241). New York: IEEE Press.
33. Melder, N., & Harwin, W. S. (2005). Force shading and bump mapping using the friction cone algorithm. In *First joint eurohaptics conference and symposium on haptic interfaces for virtual environment and teleoperator systems* (pp. 573–575). New York: IEEE Press.
34. Melder, N., Harwin, W., & Sharkey, P. (2003). Translation and rotation of multi-point contacted virtual objects. In *Proceedings of eurohaptics conference* (pp. 218–227). Trinity College Dublin and Media Lab Europe.
35. Pawar, V. M. (2013). *Exploring the influence of haptic force feedback on 3D selection strategies*. PhD thesis, University College London.
36. Rock, I., & Harris, C. S. (1967). Vision and touch. *Scientific American*, *216*(5), 96–104. doi:10.1038/scientificamerican0567-96.

37. Rock, I., & Victor, J. (1964). Vision and touch: an experimentally created conflict between the two senses. *Science, 143*(3606), 594–596.
38. Ruspini, D. C., Kolarov, K., & Khatib, O. (1997). The haptic display of complex graphical environments. In *Proceedings of the 24th annual conference on computer graphics and interactive techniques* (pp. 345–352). New York/Reading: ACM/Addison–Wesley
39. Scali, S., Wright, M., & Shillito, A. M. (2003). 3D modelling is not for WIMPs. In *Proceedings of HCI international*.
40. Schneck, C. M., & Henderson, A. (1990). Descriptive analysis of the developmental progression of grip position for pencil and crayon control in nondysfunctional children. *The American Journal of Occupational Therapy, 44*(10), 893–900.
41. Sheridan, T. (1992). *Telerobotics, automation and human supervisory control*. Cambridge: MIT Press. (Background) UR call 620.46-SHE.
42. Steinfeld, E. (1986). Hands-on architecture. http://www.access-board.gov/research/handsonarch/html/recommendations%20for%20standards.htm. December 1986. Adaptive Environments Laboratory, Department of Architecture, SUNY/Buffalo.
43. Stewart, D., & Trinkle, J. C. (2000). An implicit time-stepping scheme for rigid body dynamics with coulomb friction. In *Proceedings of the IEEE international conference on robotics and automation, ICRA'00* (Vol. 1, pp. 162–169). New York: IEEE Press.
44. Tse, B., Harwin, W., Barrow, A., Quinn, B., San Diego, J., & Cox, M. (2010). Design and development of a haptic dental training system—haptel. In *Lecture notes in computer science* (Vol. 6192, Part II, pp. 101–108).
45. Turner, A. (Ed.) (1981). *The practice of occupational therapy: an introduction to the treatment pf physical dysfunction*. London: Churchill Livingstone.
46. van den Bergen, G. (1999). A fast and robust gjk implementation for collision detection of convex objects. *Journal of Graphics Tools, 4*(2), 7–25.
47. van den Bergen, G. (2001). Proximity queries and penetration depth computation on 3d game objects. In *Game developers conference*.
48. Wall, S., & Harwin, W. (2001). Design of a multiple contact point haptic interface. In *Proceedings of eurohaptics conference* (pp. 146–148).
49. Wing, A. M., & Flanagan, J. R. (1998). Anticipating dynamic loads in handling objects. In *Proceedings of the ASME dynamic systems and control division* (Vol. 64, pp. 139–143). New York: ASME.
50. Wood, J., Magennis, M., Arias, E. F. C., Gutierrez, T., Graupp, H., & Bergamasco, M. (2003). The design and evaluation of a computer game for the blind in the GRAB haptic audio virtual environment. In *Proceedings of eurohaptics*.
51. Zilles, C., & Salisbury, K. (1995). A constraint-based god-object method for haptic display. In *IROS, international conference on intelligent robots and systems*.

Chapter 3
3D Interaction Techniques for Bimanual Haptics in Virtual Environments

Anthony Talvas, Maud Marchal, Gabriel Cirio, and Anatole Lécuyer

Abstract Bimanual haptics is a specific kind of multi-finger interaction that focuses on the use of both hands simultaneously. Several haptic devices enable bimanual haptic interaction, but they are subject to a certain number of limitations for interacting with virtual environments (VEs), such as workspace size issues or manipulation difficulties, notably with single-point interfaces. Interaction techniques exist to overcome these limitations and allow users to perform specific two-handed tasks, such as the bimanual exploration of large VEs and grasping of virtual objects. This chapter presents an overview of the current limitations in bimanual haptics and the interaction techniques developed to overcome these. Novel techniques based on the Bubble technique are more specifically presented, with a user evaluation that assesses their efficiency. These include bimanual workspace extension techniques as well as techniques to improve the grasping of virtual objects with dual single-point interfaces.

3.1 Introduction

Bimanual haptics implies the use of both hands for haptic interaction with virtual or remote environments. It is more specific than the general *multi-touch* or *multi-finger* classes of interaction in that it does not include interaction with fingers of the same hand only, but requires at least one finger of each hand. What makes two-handed interaction more specifically interesting to study is that most tasks executed in our daily lives are intrinsically bimanual [1], from simple tasks such as striking

A. Talvas (✉) · M. Marchal
IRISA/INRIA/INSA, Rennes, France
e-mail: anthony.talvas@inria.fr

M. Marchal
e-mail: maud.marchal@inria.fr

G. Cirio · A. Lécuyer
IRISA/INRIA, Rennes, France

A. Lécuyer
e-mail: anatole.lecuyer@inria.fr

I. Galiana, M. Ferre (eds.), *Multi-finger Haptic Interaction*,
Springer Series on Touch and Haptic Systems, DOI 10.1007/978-1-4471-5204-0_3,

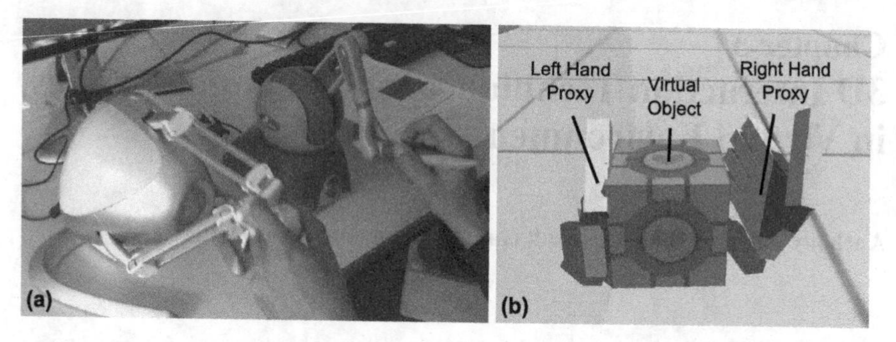

Fig. 3.1 Example of bimanual haptic interaction with a virtual environment. (**a**) Bimanual haptic setup made of two single-point devices. (**b**) Proxies and virtual object

a match to more complex ones such as playing an instrument. As such, two-handed interaction probably feels more natural to use than the one-handed kind.

This chapter focuses specifically on bimanual haptic interaction with virtual environments (VEs), which is a domain that is slowly emerging while already bearing promising applications. In the medical field, it can be used in surgery training, notably for minimally invasive surgery as illustrated by the bimanual surgical simulator interface [2] and the *da Vinci* Surgical System Simulator [3]. Another medical application is rehabilitation, as shown by a bimanual haptic desktop platform for upper-limb post-stroke rehabilitation [4]. Medicine, however, is not the only target field: non-medical applications include industrial prototyping [5] and 3D modeling software [6].

Several haptic devices are suitable for bimanual interaction, and can be subdivided into two categories: single-point and multi-finger interfaces. While the latter category allows to handle virtual objects in a similar fashion to actual hands, the former makes the task more challenging as single-point devices are represented in the virtual environment by a couple of god-objects or proxies, which is the strict minimum to allow grasping of an object (Fig. 3.1). Notably, a grasped object tends to slip from virtual hands if the contacts between proxies and objects are not strongly maintained, which shows the need for techniques to maintain more easily these contacts during grasping.

Another issue is that, regardless of which category they belong to, most of the bimanual haptic interfaces have small workspaces, which is a strong limitation when a user wants to carry out tasks as simple as picking an object and moving it around in a large VE. Several techniques to extend the workspaces of haptic devices within a VE were proposed. A notable solution is the use of metaphors, *i.e.* analogies that help the user execute a given task in an intuitive manner, such as the *Bubble* technique [7] which proved to be suitable for simultaneous grasping of objects and exploration in a VE with a bimanual whole-hand haptic interface [8]. More recently, novel metaphors adapted this *Bubble* technique to any couple of 3DoF single-point haptic interfaces [9], also allowing grasping and navigation at the same time.

These novel techniques include workspace increasing techniques for navigating in large VEs with two haptic devices, as well as grasping techniques to simplify the manipulations of objects in such environments. This chapter first presents the current limitations of bimanual haptics, and related work on bimanual haptic interaction techniques. The aforementioned novel interactive techniques based on the *Bubble* technique are then presented, followed by a user evaluation. Finally, perspectives of bimanual haptic interaction techniques are presented, before reaching a conclusion.

3.2 Related Work

The hardware and software solutions that allow bimanual haptic interaction with VEs are currently subject to some limitations, which will be overviewed in this section. Interaction techniques and metaphors, which combine hardware, software and knowledge of human cognition to allow a user to perform a given task, can notably help to overcome these limitations. Very few genuinely two-handed haptic interactive techniques were developed to this date, but some techniques from unimanual haptic interaction and two-handed interaction in 3D environments, not necessarily haptic, can be adapted for a bimanual haptic case. This section will provide a global overview of the different interactive techniques suitable for bimanual haptic interaction with VEs.

3.2.1 Hardware and Software Limitations for Bimanual Haptics

Several haptic devices allow bimanual interaction with VEs. However, they share the same drawback of having workspaces fairly limited in size, thus preventing the interaction with large VEs without using techniques to increase these workspaces. This problem can be handled at the hardware layer, by improving the existing hardware or by using additional devices.

A straightforward and general approach consists in increasing the workspace provided by the devices, either by using large serial linkages, string-based devices with large frames, or redundant degrees of freedom (DoF) [10]. For unimanual interaction, human-scale devices could be developed using that approach. However, as far as bimanual devices are concerned, such scales could not be obtained and the highest reach that was achieved so far was that of human arms, in the cases of the DLR interface [5] and the Haptic Workstation [8].

Another solution for providing larger workspaces is the use of mobile haptic interfaces, *i.e.* haptic devices fixed on a mobile robot. Bimanual examples of these are the Mobile Haptic Grasper [11] and VISHARD7-based mobile interface [12]. However, even though they potentially provide an infinite planar workspace [13], these devices are still limited in vertical reach.

Other hardware approaches address the workspace issue by providing additional DoF to the user to handle navigation in the VE. For instance, a 3DoF foot pedal was used for controlling the motion of a two-armed robot in a remote environment [14]. The latter approach may be the only one to provide an infinite workspace in all directions, but is fully dependant on the availability of such pieces of hardware.

As far as manipulation capabilities of haptic devices are concerned, whether unimanual or bimanual, those devices can be divided into two main classes. The first category is that of single-point interfaces, such as the SPIDAR G&G [15], the DLR bimanual haptic interface [5] or the more widespread PHANToM series. The second category is that of multi-fingered interfaces, such as the MasterFinger-2 [16], SPIDAR-8 [17] and Bimanual HIRO [18], which enable interaction through 4, 8 and 10 fingertips respectively. Within this category of multi-finger devices, the Haptic Workstation is a special case, providing whole-hand input through the CyberGlove, haptic feedback to the fingers with the CyberGrasp, and to the palms of both hands with the CyberForce [8].

Multi-finger interfaces are well suited for grasping, as more contact points with objects means better handling of the said objects: it was shown that 3 contact points with friction, or 7 contact points without friction, were the minimum requirements for form closure of an object, *i.e.* the ability to fully restrain an object [19]. However, manipulating virtual objects with two single-point haptic interfaces is a different matter. It was shown that 2 contact points could allow full manipulation of an object, but only if the contacts are soft-finger, meaning being able to support moments around the contact normal [20]. Even so, handling virtual objects with single-point devices is very much like attempting to handle an object with the fingertips of index fingers, and as such a challenging task.

Concerning the software layers that link the aforementioned devices to the VEs, there are various haptic APIs that allow to do so, which can be divided into two categories. Device-specific APIs usually grant low-level access to one model or series of interfaces, generally allowing the use of the same device in a dual way. However, it is not possible to conjointly use devices of two different manufacturers using only one of these APIs, but it is possible to combine the respective APIs of the two devices. For instance, HDAL SDK [21] and OpenHaptics [22] can be used to operate conjointly a Novint Falcon and a PHANToM device.

This process can be bothersome, and thus generic APIs, like CHAI3D [23], were developed to support a wider range of devices, not limited to a single manufacturer. Some of them have clearly shown their bimanual capabilities, such as H3DAPI [24], which was used for a bimanual haptic 3D modeling software [6], and Haptik Library [25], which was part of the software architecture of the Mobile Haptic Grasper [11]. Both cases used twice the same device, but these APIs can also be used with two different devices ; notably, examples shown later in this chapter used Haptik Library to handle a Novint Falcon and a PHANToM Omni.

However, while these APIs support the use of two devices at the same time, they do not truly support bimanual capabilities *per se*. There are for instance no solutions to reduce the impact on the user of different workspace sizes, output forces, or degrees of freedom in each hand. Similarly, while methods exist to prevent collision between the hands of the user and the two devices [18], they are not yet implemented in these APIs either.

3.2.2 Interaction Techniques for Bimanual Haptics

The hardware approaches for increasing the workspaces of bimanual haptic hardware do effectively allow the exploration of large VEs to some extent. However, on the one hand, most of these techniques do not increase those workspaces infinitely, and on the other hand, such devices are not necessarily widespread. This leads to the use of interaction techniques for handling the extension of these workspaces. They have the advantage of being generic and applicable to any haptic device available to the user with no further requirements, although the majority of currently existing techniques are not specific to two-handed haptics.

A first technique consists in applying a scaling factor to match the real workspace provided by the haptic devices with a virtual volume defined in the VE [26], although reducing the accuracy of motions in the virtual space. Another approach is the *clutching* technique, which consists in holding down a button to temporarily interrupt the coupling between the device and the proxy while the user recenters the device. The *Dual Shell* method is an extension of this technique, that automatically handles the clutching when predefined boundaries are reached, without requiring the potentially counterintuitive manipulation of a button [27]. The use of rate control was also proposed to control the velocity of the virtual proxy through the position of the haptic device [28]. This technique infinitely increases the workspace in all directions, although seeming to be far from intuitive.

A last interaction technique is the *Bubble* metaphor, which is originally a unimanual haptic workspace extension technique [7]. It uses direct control of the position of the proxy in the VE while the device remains inside predefined spherical boundaries (defined as the *bubble*). When the device moves outside this *bubble*, the latter is translated throughout the virtual environment, with a small elastic force that gives the feel of the surface of the *bubble* to the user. For a unimanual painting task, a user experiment showed that this technique was faster, more accurate and more appreciated than the scaling factor and clutching techniques.

The bimanual case was not yet throughly investigated with the *Bubble* technique, though it was used for bimanual interaction with complex VEs through the Haptic Workstation [8]. This implementation allowed users to translate and rotate the camera by moving both hands outside the bubble in the same direction. It showed to be efficient for simultaneous navigation and manipulation with this specific device, as it allows interaction with the palms and fingers of both hands. However, it remains difficult to use with single-point interfaces, as picked objects tend to be frequently dropped during the translations of the virtual workspace through rate control. This is particularly true when using two different interfaces with physical workspaces of different size and shape, a scenario that has been scarcely studied.

Aside the question of how to explore large VEs with bimanual interfaces, next comes the matter of manipulating virtual objects with these devices. Bimanual haptic manipulation techniques vary greatly depending on the task that is to be executed. For instance, the multi-modal mesh manipulation system [6], a bimanual 3D modeling software, uses a tool-object metaphor. It can assign a grabbing task to the non-dominant hand and different manipulation tools to the dominant hand, or a tool in each hand for bimanual manipulation of the mesh (stretching, folding, tearing, *etc.*). Similarly, the Bimanual Haptic Simulator for medical training [29] assigns a palpation task to the non-dominant hand and a needle insertion task to the dominant hand.

However, while these techniques work for a specific task, the question of how to switch between different tasks and techniques remains open. In the context of 3D interaction techniques, different methods were proposed to apply transitions between different tools, or subtasks, proposed to the user in the case of the Responsive Workbench [30]. They defined explicit transitions, like picking up a tool in a toolbox, with a default behaviour when no tool has been picked up yet, that should be specific of the application. Another example of explicit transition, more practical in the sense that it does not require several movements between the workspace and the toolbox, is the use of gestures with the hands. Implicit transitions were also defined, in which a switch from an subtask to another occurs in a seamless way, almost imperceptible to the user. An example of this is the switch from a unimanual grabbing technique to a bimanual grab-and-twirl technique, that occurs naturally as the user reaches in with the other hand to help the manipulation.

There is a more general two-handed manipulation task that several studies focused on, which is the grasping of arbitrary virtual objects. A technique was proposed to detect when a user attempts to grasp an object, based on a study on the segmentation of a grasping task with multi-finger interfaces [31]. The study distinguished three major steps with specific forces applied on the grasped object: approach (no force applied on the object), gripping (a horizontal force being applied), and lifting (a vertical component being added). The information on the forces applied by each finger on a virtual object can thus allow the controller to detect grasping and simulate it accordingly.

Beyond detecting grasping, a straightforward interaction technique for handling bimanual multi-finger manipulation of virtual objects is the use of virtual hands, such as spring-damper hands using either fully rigid phalanxes [8, 32] or a rigid skeleton with deformable fingerpads [33, 34], but also using fully deformable hands [35]. A generalized god-object method was notably proposed to simulate realistically the behaviour of virtual hands by avoiding the penetration of objects and other artifacts [36]. However, as mentioned earlier, manipulating objects with two single-point interfaces is intrinsically more challenging: to the best of our knowledge, no manipulation technique tackles this problem.

3.3 Interaction Techniques for Improving Bimanual Interaction with Single-Point Haptic Interfaces

Bimanual interaction with VEs using single-point haptic interfaces implies interacting with virtual objects using one contact point per hand. More than just allowing to perform single-handed tasks in a dual manner, this kind of interaction gives access to specifically two-handed tasks, such as grasping. Picking, lifting and carrying an object with two contact points isn't specific to bimanual interaction, though, as it can also be done with two fingers of the same hand. However, picking a virtual object two-handedly with single-point devices is very much like attempting to pick a real object with only one fingertip of each hand, which is more difficult than doing so with two fingertips of the same hand. The question of how to reduce that difficulty has, to this date, been scarcely studied. Thus, a set of bimanual interaction techniques was proposed for improving the exploration of large VEs with single-point interfaces, as well as the manipulation of objects with two 3DoF haptic devices represented by simple proxies [9].

This section will present these techniques, which include two haptic exploration techniques, a grasping detection method, and two haptic manipulation techniques. The exploration techniques are the *double bubble*, which allows free motion with both hands in a VE, and a viewport adaptation method that maintains both virtual proxies on screen. The grasping detection method allows to determine when a user attempts to effectively pick an object between both hands. The first manipulation techniques is the *magnetic pinch*, which uses a simulated spring to keep the virtual proxies from dropping a picked object. Finally, the second manipulation technique is the *joint control*, which facilitates grasping and manipulation of virtual objects by keeping both hands from being in different control modes (position or rate) or having different scale factors when interacting with the same object.

3.3.1 Double Bubble

In the *double bubble* technique, the workspace of each haptic device is defined by two areas, each associated to a control mode (Fig. 3.2). An inner area controls the proxy directly in position, and an outer area, starting at the boundaries of the inner area and extending up to the physical limits of the device, controls the virtual workspace in speed within the VE (Fig. 3.3).

Besides using two interfaces instead of one, two major differences separate the *double bubble* from the previously mentioned *bubble* technique. The first difference is the use of a rectangular parallelepiped for the boundaries of the *bubbles* instead of a sphere, to better fit the physical workspaces of the devices. We can notably think of PHANToM devices which have a higher width than their height or depth. The second difference is the presence of a visual feedback added to the haptic feedback when the devices leave the boundaries, in the form of a trail behind the rate-controlled proxies (Fig. 3.3).

An issue that was observed with several users attempting to use the technique was that they got confused whenever the right proxy was on the left of the screen and vice versa. A way to prevent this is to simulate an invisible "separation plane" that prevents the centers of the bubbles from crossing, by negating the horizontal component of the *bubble* velocities when such situation is about to happen. Given the left *bubble* of center $\vec{l} = (l_x, l_y, l_z)$ and its displacement at the next simulation

Fig. 3.2 Approximate *bubble* and physical workspace sizes of a Novint Falcon and a PHANToM Omni. *Inner rectangles* represent the *bubbles*, while *outer polygons* represent the physical workspaces (mechanical stops of the devices)

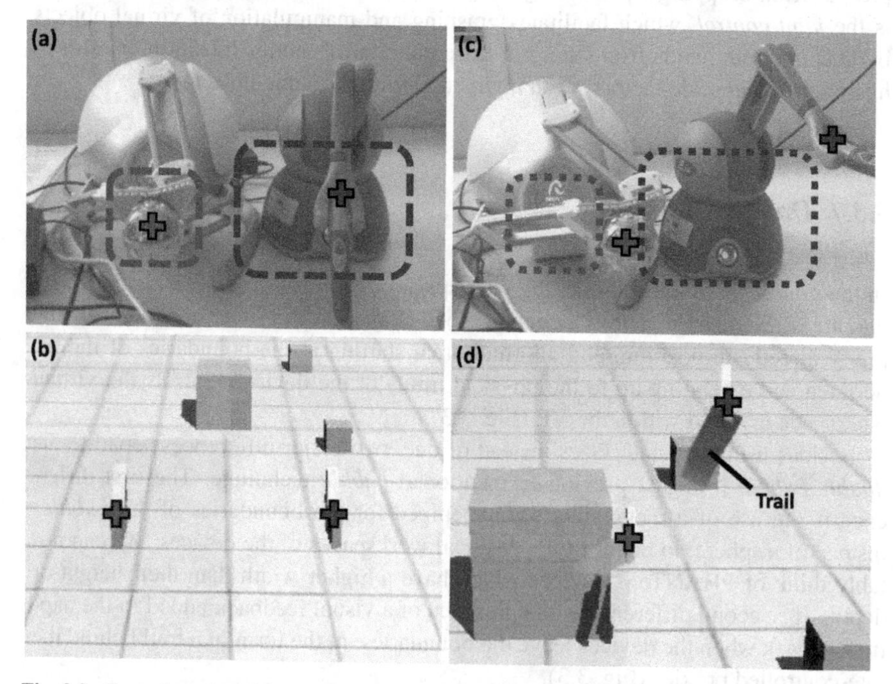

Fig. 3.3 Control modes of the *double bubble*. (**a–b**) *Devices inside* the *bubbles*: position control. (**c–d**) *Devices outside* the *bubbles*: rate control

step $\vec{d} = (d_x, d_y, d_z)$, and the right *bubble* of center $\vec{r} = (r_x, r_y, r_z)$, the constraint is applied following (3.1). The same constraint is applied for the right bubble.

$$\vec{d} = (r_x - l_x, d_y, d_z) \quad \text{if } l_x + d_x > r_x. \tag{3.1}$$

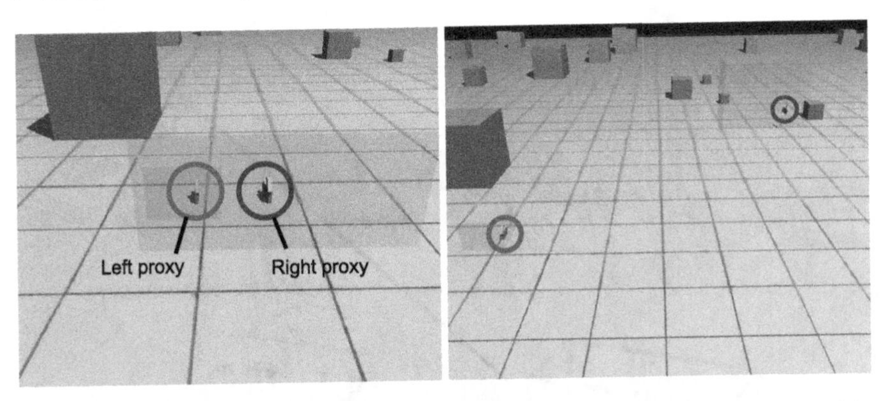

Fig. 3.4 Result of the viewport adaptation with different relative positioning of the proxies (*circled*)

3.3.2 Viewport Adaptation

Since each device is attached to a *bubble* independent from the other, a method is required to keep both proxies on the screen, as they can move infinitely in completely opposite directions. Thus, a method was developed to ensure that both *bubbles* remain in view no matter the distance between them (Fig. 3.4). This is accomplished by setting the distance of the camera to the center of the scene to a value proportional to the distance between the leftmost border of the left *bubble* and the rightmost border of the right *bubble*, plus an arbitrary margin (Fig. 3.5). Given the left *bubble* of center $\vec{l} = (l_x, l_y, l_z)$ and width w_l, and the right *bubble* of center $\vec{r} = (r_x, r_y, r_z)$ and width w_r, the position of the camera is computed following (3.2)–(3.4).

The center of scene $\vec{s}$ is first computed from both workspace centers following:

$$\vec{s} = \left(\frac{l_x + r_x}{2}, \frac{l_y + r_y}{2}, \max(l_z, r_z) \right). \tag{3.2}$$

The width of the displayed scene w_s is then computed from the widths of both workspaces w_l and w_r, as well as an arbitrary margin m that ensures that the virtual workspace boundaries do not leave the borders of the screen:

$$w_s = \sqrt{(r_x - l_x)^2 + (r_y - l_y)^2} + w_l/2 + w_r/2 + 2m. \tag{3.3}$$

Finally, the position of the camera $\vec{c}$ is computed following:

$$\vec{c} = \vec{s} + w_s \cdot d \cdot \vec{a}. \tag{3.4}$$

where d is a scalar that depends on the camera field of view, and $\vec{a}$ is an arbitrary unit vector that determines the angle from which the scene is displayed.

A limitation of this method is that it does not allow rotations of the viewport. A way to handle these rotations is to use the "separation plane" mentioned in the previous section, by translating the x component of the displacement of the *bubble* into an angular velocity that will be applied to the vector $\vec{a}$ (Fig. 3.6). The displacement of both proxies can be used, making the viewport rotation faster if both hands push in opposite directions simultaneously.

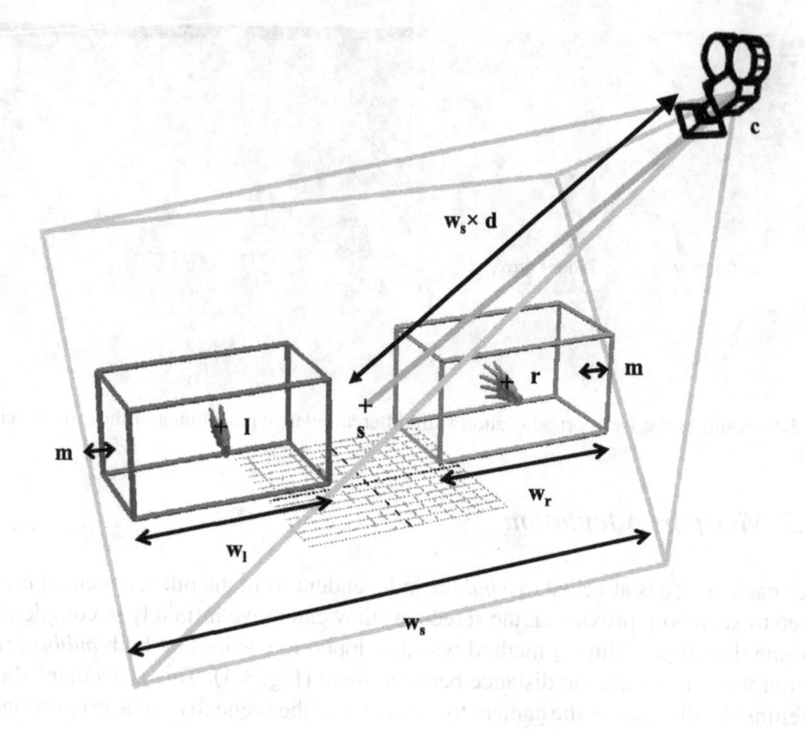

Fig. 3.5 Computation of the camera position for viewport adaptation

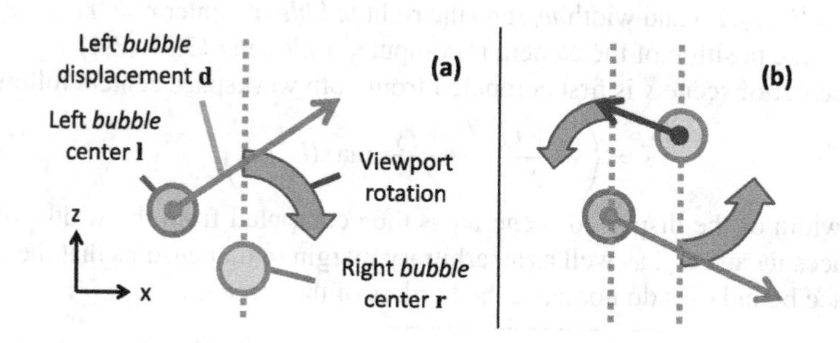

Fig. 3.6 Rotation of the viewport by "pushing" on the separation plane with the proxies. (a) One-handed case. (b) Two-handed case

3.3.3 Grasping Detection

A grasping detection method is required in order to detect when a user is actually attempting to pick an object and not simply touching it. Three conditions are considered to determine whether both hands are grasping an object or not, according to the contact normals, the contact forces, and the relative position of both hands (Fig. 3.7):

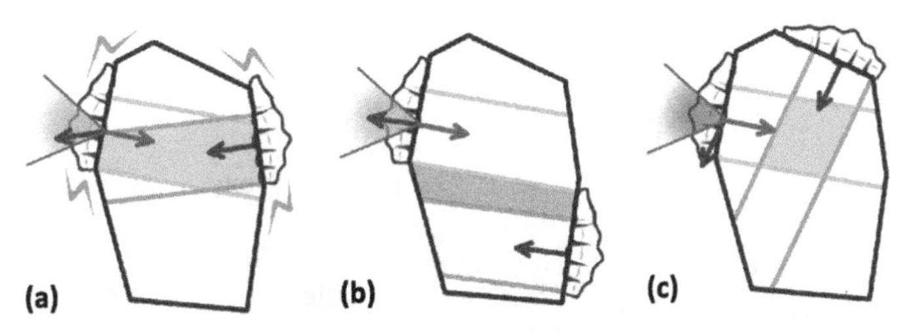

Fig. 3.7 Different cases of dual contact with a virtual object, case (**a**) being appropriate for grasping and cases (**b–c**) not being as such: (**a**) Normals nearly colinear and hands face-to-face, (**b**) Hands not in front of each other, (**c**) Normals far from colinearity

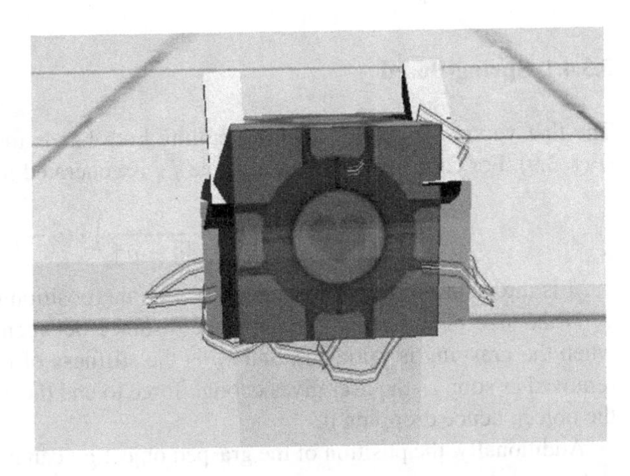

Fig. 3.8 Visual feedback of the *magnetic pinch*, symbolized by *bolts* between virtual proxies and picked object

1. The angle between the contact normals must be under a certain threshold.
2. Both contact forces must exceed a threshold in order to discriminate simple contacts with an object from a true intent of grasping the object.
3. Two cylinders projected from both proxies following the contact normal must intersect. Their radii match the approximate sizes of the proxies by default, but can be tuned to make the detection more or less sensitive.

3.3.4 Magnetic Pinch

Once a grasping situation is detected, the *magnetic pinch* takes effect, which "magnetizes" both hands to the picked object to prevent unintentional drops from happening. A visual feedback is also added to the haptic feedback in the form of red bolts, emphasizing the activation of the technique to the user (Fig. 3.8). Two implementations were considered for the magnetic pinch: one based on springs and the other based on constraints.

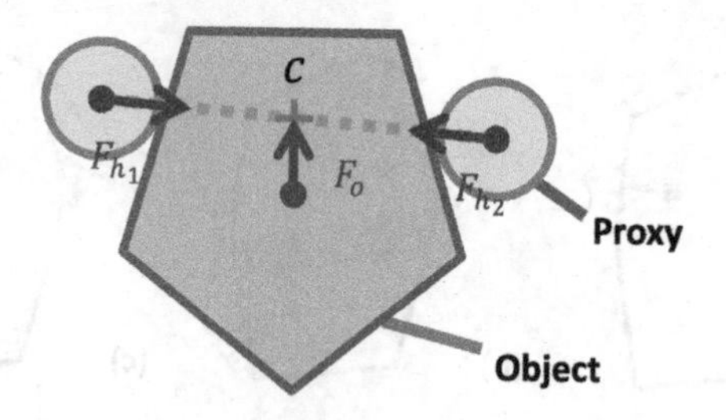

Fig. 3.9 Forces applied by the spring-based approach of the *magnetic pinch*. F_{h_1} and F_{h_2} are the forces applied on the centers of mass of both proxies to pull them towards each other. F_o is the force applied on the center of mass of the picked object to pull it towards c, the middle point between both virtual proxies

3.3.4.1 Spring-Based

The first version simulates a spring pulling both hands towards the picked object (Fig. 3.9). For each haptic device, a force $\vec{F_h}$ is generated following:

$$\vec{F_h} = -k_h \cdot \left(1 - \frac{g_s}{\|\vec{o} - \vec{p}\|}\right) \cdot (\vec{o} - \vec{p}). \tag{3.5}$$

$\vec{p}$ is the position of the first interface, $\vec{o}$ is the position of the second interface, g_s is the size of the grasped object (the distance between the two contact points when the grasping is initiated), and k_h is the stiffness of the spring. The spring is removed as soon as the user gives enough force to end the contact of the hands with the object, hence dropping it.

Additionally, the position of the grasped object $\vec{g_p}$ can be constrained to the central point between the positions of the two virtual proxies $\vec{l}$ and $\vec{r}$, further reducing the risk of unwanted drops. For this, another spring is used, of stiffness k_o with a force $\vec{F_o}$ following:

$$\vec{F_o} = -k_o \cdot \left(\frac{\vec{l} + \vec{r}}{2} - \vec{g_p}\right). \tag{3.6}$$

This approach successfully stabilizes the grasping of light virtual objects. However, lifting heavier objects can still be problematic with this approach, as well as correctly handling the rotations of the picked object.

3.3.4.2 Constraint-Based

The second approach for the *magnetic pinch* constrains the position and orientation of both proxies relative to the picked object. Both constraints are removed as soon as at least one proxy provides an outwards force which value exceeds an arbitrary threshold. This threshold determines the weight of the objects that can be picked with the magnet effect: the higher the value, the heavier the object that can be lifted. However, higher values also imply that strong forces must be applied to release even

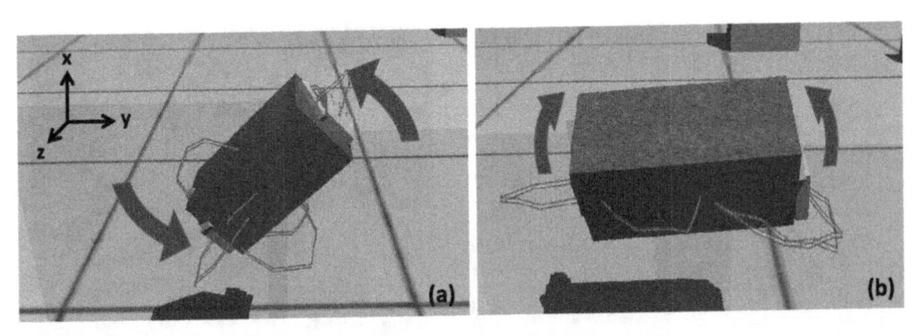

Fig. 3.10 Rotation of a virtual object using the constraint-based approach of the *magnetic pinch*: (**a**) Rotation around the z axis using the relative position between both hands. (**b**) Rotation around the y axis using the torque of the right device

lighter objects, which can give an unnatural feel of being abnormally "glued" to the object in these cases. This effect could be reduced by dynamically modulating the threshold with the weight of the picked object.

Not only does this approach allow to carry heavier virtual objects, but it also allows the rotation of these objects on two degrees of freedom, even with 3DoF devices, by using the relative position between both proxies (Fig. 3.10a). In addition, if at least one of the two devices has 6DoF sensing (possibly underactuated), then it is possible to handle the third rotational degree of freedom of the object (Fig. 3.10b).

3.3.5 Joint Control

The *double bubble* metaphor may introduce a difference in control modes and/or scaling factors when activated (Fig. 3.11). In order to reduce the impact of these differences when picking and placing a virtual object, the notion of *joint control* was introduced. During a grasping situation, both devices use a common control/display ratio (average of both) and common bubble size (minimal dimensions), and enter rate control simultaneously when at least one device leaves its *bubble*. This technique allows easier exploration of a VE while holding an object between virtual hands. It is especially helpful when the proxies are controlled by two different haptic devices, but is also of use with same devices.

3.4 Evaluation

An experiment assessed the efficiency of the previously mentioned techniques, involving a simple pick-and-place task, where users had to pick a cube and place it at a given position. The *double bubble* technique was compared to the *clutching* technique for workspace extension, and the benefits of the *magnetic pinch* and *joint control* were also measured for grasping facilitation.

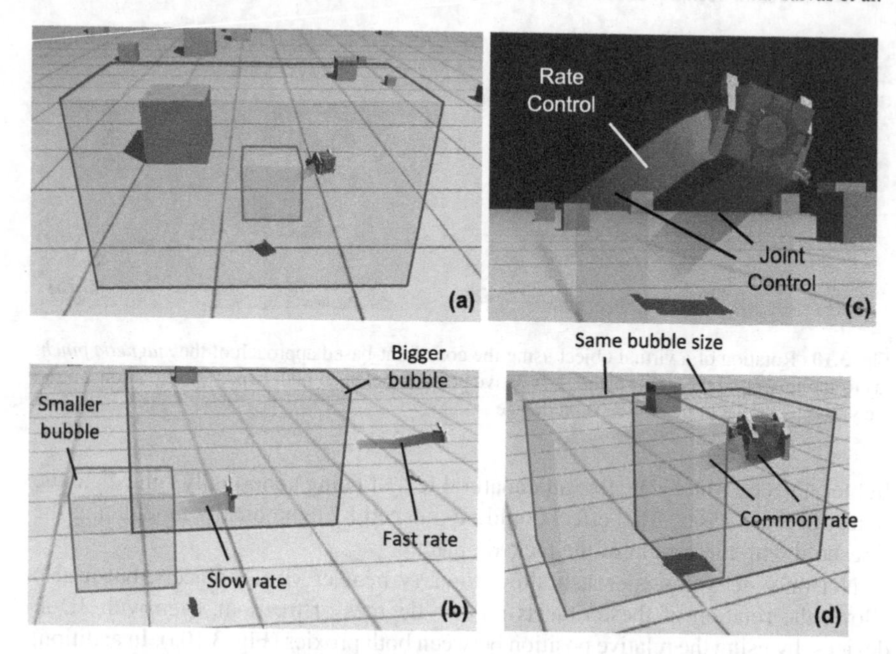

Fig. 3.11 Illustration of *joint control*. (**a**) Carrying an object without *joint control*; case where the *left*, smaller *bubble* is in rate control and not the *right*, bigger *bubble*. (**b**) Difference in *bubble* size and workspace translation speed without *joint control*. (**c–d**) Carrying an object with *joint control*

3.4.1 Method

3.4.1.1 Population

Thirteen participants (2 females and 11 males) aged from 20 to 26 (mean = 22.8, sd = 1.7) performed the experiment. None of the participants had any known perception disorder. All participants were naïve with respect to the proposed techniques, as well as to the experimental setup and the purpose of the experiment.

3.4.1.2 Experimental Apparatus

The participants were seated at 1 m in front of a 24 inch widescreen monitor. The experiment was conducted using two different haptic interfaces. The participants manipulated a Falcon (Novint Technologies Inc., Albuquerque, New Mexico, USA) in their left hand, and a PHANToM Omni (Sensable Technologies, Wilmington, Massachusetts, USA) in their right hand, both placed in front of the screen as shown in Fig. 3.12. Visual feedback was rendered at a refresh rate of 50 Hz, while the haptic rendering rate was 1,000 Hz. Physical simulation was performed using Nvidia PhysX at a rate of 1,000 Hz to match the update frequency of the haptic loop. A virtual coupling mechanism

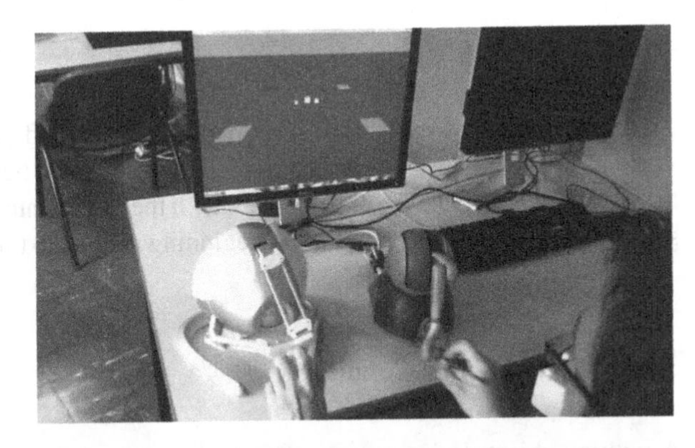

Fig. 3.12 Apparatus used in the experiment

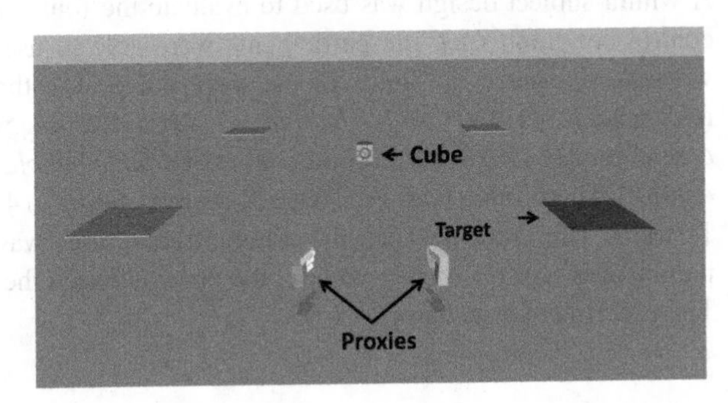

Fig. 3.13 Virtual environment used in the experiment

[37] was used between the haptic interfaces and the virtual proxies by simulating a spring-damper system between each haptic device and its corresponding proxy.

3.4.1.3 Virtual Environment

The VE was composed of a 100 m-wide ground plane with four potential target planes, of 1 m of width, placed at the corners of a 6 m-wide square around the center of the VE. The target plane of each trial was colored in red, and the other planes were colored in white. The cube to be manipulated had a width of 30 cm and a mass of 3 g, and was placed at the center of the VE. The proxies controlled by each haptic device were physically represented by cubes of 20 cm of width, and were positioned 2 m away from each other and 5 m away from the central cube at the start of each trial. The sizes of the cube, targets and proxies were chosen so as to match a real life task. The distances between them were chosen so as to require both proxies to get out of the workspace borders to reach the cube and targets. The proxy controlled by the left device was visually represented by a blue left hand, and the right proxy was represented by a green right hand. Figure 3.13 shows the scene as displayed at the beginning of a trial.

3.4.1.4 Procedure

At the beginning of each trial, both haptic devices and proxies were set to their starting positions. The subject had to pick the cube from both sides, carry it towards the red target and make the cube contact with the target, thus ending the trial. A black screen warned the subject about the beginning of the next trial.

3.4.1.5 Experimental Conditions

A within-subject design was used to evaluate the four different conditions. In the control condition *Ctrl*, the participants were able to use the *clutching* technique when they reached the limits of the workspaces. The three other conditions corresponded to: (1) *DB* (*double bubble*), (2) *MP* (*clutching* with *magnetic pinch/joint control*) and (3) *DB*+*MP* (a combination of *double bubble* and *magnetic pinch/joint control*). There was a total of 176 trials, corresponding to 4 conditions × 4 targets × 11 trials. The order between the different conditions was counterbalanced across participants, and for each condition, the order between the targets was randomized. The experiment lasted around 1 hour.

3.4.1.6 Collected Data

For each trial and each participant, the completion time and number of drops were recorded. The completion time is the time elapsed between the moment the proxies leave their starting positions and the moment the cube touches its target plane. The number of drops is the number of hits recorded between the cube and any part of the ground plane that is not the target plane. At the end of the experiment, participants had to complete a subjective questionnaire in which they had to grade the different techniques according to different criteria. The participants could rate the criteria from 1 (very bad) to 7 (very good). The different criteria were: (1) Global appreciation, (2) Efficiency, (3) Learning, (4) Usability, (5) Fatigue, and (6) Realism.

The following questions were also asked to the participants:

- Have you encountered difficulties during the experiment?
- Did you perceive differences between the different conditions?
- What manipulation strategies did you use for the different conditions?
- Which condition do you prefer and why?
- If you were to perform a more complicated manipulation task such as manipulating with obstacles or realizing assembly tasks, which condition would you have preferred?
- Do you have remarks, suggestions?

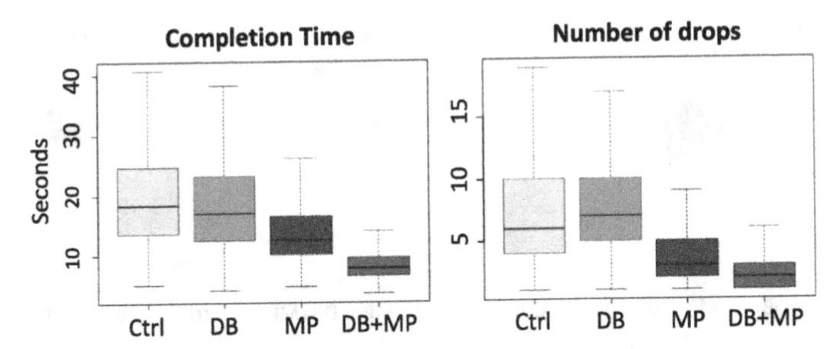

Fig. 3.14 Box plots of the completion times and number of drops for all conditions. They are delimited by the quartile (25 % quantile and 75 % quantile) of the distribution of the condition over the individuals. The median is represented for each trial

3.4.2 Experiment Results

Concerning the data, we performed a Shapiro test that rejected the normality hypothesis on the data distribution. Thus, we used a non-parametric Friedman test for differences among the different projections. Post-hoc comparisons were performed using Wilcoxon signed-rank tests with a threshold of 0.05 for significance. Reported p-values are adjusted for multiple comparisons.

3.4.2.1 Completion Time and Number of Drops

A statistical analysis was conducted from the completion time data collected during the experiment. The results are displayed on Fig. 3.14. For each participant, statistics (mean M, standard deviation *SD*) were computed on the 44 trials in each condition. A Friedman test on the completion time (in seconds) revealed a significant effect of the technique ($\chi^2 = 27.66$, $p < 0.001$). Follow-up post-hoc analysis revealed that completion time in both the *MP* ($M = 14.16$, $SD = 7.14$) and *DB*/*MP* ($M = 8.43$, $SD = 2.91$) conditions were significantly shorter that in the control ($M = 21.41$, $SD = 13.19$) and *DB* ($M = 20.06$, $SD = 14.63$) conditions ($p < 0.001$ in all cases), and that the $DB + MP$ condition led to significantly shorter times than the *MP* condition as well ($p < 0.001$).

Similarly, a statistical analysis was conducted on the number of drops for all trials of each participant. A Friedman test showed a significant effect of the technique ($\chi^2 = 25.52$, $p < 0.001$). Post-hoc analysis showed that the *MP* ($M = 4.22$, $SD = 9.45$) and *DB*/*MP* ($M = 2.36$, $SD = 2.33$) conditions led to significatively less drops than the control ($M = 7.88$, $SD = 6.37$) and *DB* ($M = 8.79$, $SD = 6.77$) conditions ($p < 0.001$ in all cases).

3.4.2.2 Subjective Questionnaire

The results for the subjective questionnaire are displayed on Fig. 3.15. A Friedman test was performed to analyse the answers of the participants. The reported p-values

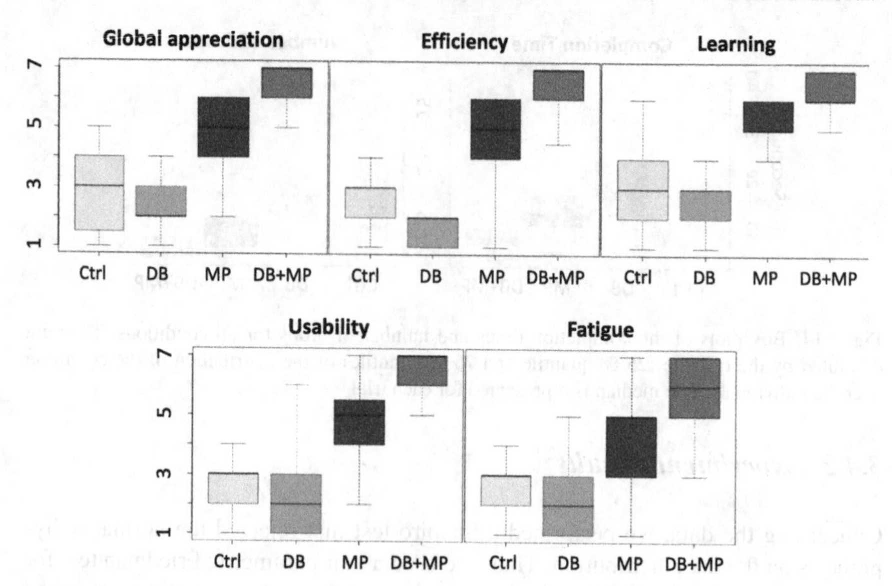

Fig. 3.15 Box plots of the subjective ratings for the significative criteria, for all conditions. They are delimited by the quartile (25 % quantile and 75 % quantile) of the distribution of the condition over the individuals. The median is represented for each trial

were adjusted for multiple comparisons (alpha-level $p = 0.05$). A significant effect was found for 5 criteria: Global appreciation ($\chi^2 = 4.62$, $p < 0.001$), Efficiency ($\chi^2 = 4.92$, $p < 0.001$), Learning easiness ($\chi^2 = 4.50$, $p < 0.001$), Use easiness ($\chi^2 = 4.80$, $p < 0.001$) and Fatigue ($\chi^2 = 4.46$, $p < 0.001$). No significant effect was found for the Realism criteria.

Post-hoc analysis showed that the $DB + MP$ condition was preferred to both the control and *DB* for all criteria: Global appreciation ($p < 0.001$ and $p < 0.001$ respectively), Efficiency ($p < 0.001$ and $p < 0.001$), Learning ($p < 0.001$ and $p < 0.001$), Usability ($p < 0.001$ and $p < 0.001$) and Fatigue ($p < 0.001$ and $p < 0.001$). The *MP* condition was also preferred over the control and *DB* for 3 criteria: Global appreciation ($p = 0.029$ and $p = 0.028$), Learning ($p = 0.032$ and $p = 0.009$) and Usability ($p = 0.027$ and $p = 0.008$), plus a fourth criteria for the *DB*: Efficiency ($p = 0.020$).

All participants perceived well the differences between the four conditions. Most participants stated that they preferred the $DB + MP$ condition over all the others, reported as faster and easier to use, as well as more efficient and less tiring. They also mentioned this condition as being their condition of choice for more complex manipulation tasks. One participant, however, preferred the *MP* condition, mentioning its behaviour was close to that of a mouse. For the conditions without grasping techniques, some participants had a tendency to throw the cube towards the target rather than actually putting it down with both hands like they did with the grasping techniques. It was also noted that the lack of magnetic pinch sometimes led the user to apply too strong forces on the cube, causing it to drop. A few participants also noted the clutching technique as being overall difficult, while others mentioned the difficulty of having two different interfaces regardless of the technique.

3.4.3 Discussion

The conducted experiment showed that the *magnetic pinch* and *joint control* improved performances and subjective appreciation for a pick-and-place task over the *double bubble* and *clutching* navigation techniques, while the combination of these techniques with the *double bubble* led to the best results.

The *double bubble*, used alone, performed as good as the *clutching* technique without outperforming it, in terms of completion time, drop rate, and subjective appreciation. The technique allows to translate the workspace in a VE in a smoother way than the *clutching* technique, by removing the need to move the devices back and forth several times.

The experiment showed that the *magnetic pinch* and *joint control* significantly reduced completion times and dropping rates compared to the conditions that did not use them. In addition, the subjective appreciation also favored the conditions which used these techniques over those that did not, globally as well as for learning and usability. These results strongly indicate that the *magnetic pinch* and *joint control* techniques, by stabilizing the grasping of a virtual object with virtual proxies, are efficient for facilitating pick-and-place tasks. Additionally, while the *magnetic pinch* inherently adds an unrealistic behaviour through the magnetic attraction, it does not seem to hinder the global realism of the scene, as no significant difference in the participants perception of realism was reported for the different conditions.

The *double bubble* showed its full potential when used jointly with the *magnetic pinch* and *joint control*, outperforming the combination of the latter techniques with *clutching*. The *double bubble* allows users to perform the task in a simpler and faster way than the *clutching* technique, which imposes frequent stops of both proxies to recenter the two haptic devices.

3.5 Perspectives

Haptic interaction with virtual environments using two hands is a topic that is being increasingly investigated but is still far from reaching its full potential. Interaction techniques dedicated to that field are not numerous, as in a majority of cases unimanual techniques are used in a dual way rather than developing actual bimanual techniques and metaphors. Notably, this means that the same tasks and possibilities are given to both hands, not taking into account the intrinsic differences between them and their ability to perform asymmetric tasks. Exceptions are studies that focus on specific tasks such as medical training [29] or 3D modeling [6], which assign a different tool to each hand prior to the beginning of the task, with possibly two-handed interactions between these tools.

All haptic devices suited for bimanual interaction have fairly limited physical workspaces, raising the need for interaction techniques to extend them. Most of the existing haptic exploration techniques that allow to do this are unimanual, notably the *clutching* and *Bubble* techniques. The *Double Bubble* metaphor extended the latter to a bimanual use, allowing independent translation of the virtual workspaces of both devices within a large virtual environment through hybrid position/rate control modes. While the *clutching* technique leads to "jerky" motions with the need to make frequent stops when the boundaries of the workspaces are reached, the *double bubble* allows much smoother motions in virtual environments. These techniques only allow translational movements originally, but two approaches allow to rotate the viewport as well: the bimanual implementation of the original *bubble* technique (at the expense of lateral translations), as well as the separation plane of the *double bubble*.

While multi-finger interfaces enable good manipulations of virtual objects through their multiple contacts, it is not the case for single-point devices. The grasping of objects with these devices can be tedious, thus raising the need for interaction techniques such as the *magnetic pinch* to simplify the lifting and carrying of objects with only two contact points. Applying a small magnet effect does help for maintaining the grip of the object, although some issues remain. Notably, holding an object with two haptic devices allows to translate the said object in all directions, as well as rotate it using the relative position between the devices, but there is a degree of freedom that cannot be controlled at all with 3DoF devices. Another point worth noting is that the grasping detection doesn't take into account the mass of the picked objects, while it was shown that the force required to grip an object without slipping depends on mass, stiffness and friction parameters [38]. Also, while this technique was developed for single-point interfaces, its relevance for multi-finger interaction and the question of how to adapt it to that case remain to be studied. Finally, the principles work perfectly for rigid bodies, but whether they could work and/or be adapted for interaction with deformable bodies is yet to be explored.

It was shown that more issues arise when exploration and grasping happen simultaneously, mostly with single-point devices. With *clutching*, drops tend to happen during the temporary decoupling between devices and proxies, while with the *double bubble*, the differences of velocities in rate control as well as times of activation of rate control can lead to unwanted drops as well. Thus, techniques like the *joint control* are required to handle these two classes of haptic interaction simultaneously. However, there are still several issues to be addressed, like the rotation of the viewport with an object in hand.

3.6 Conclusion

This chapter focused on haptic interaction with VEs using both hands, by overviewing several interaction techniques suited for two-handed haptics. Among them, workspace extension techniques such as the *clutching* and *double bubble* allow bimanual haptic exploration of large VEs, the latter leading to smoother motions than the former. While multi-finger interfaces allowed good handling of virtual objects through their multiple contact points, single-point devices strongly benefit from interaction techniques such as the *magnetic pinch* that maintain the contacts between virtual proxies and picked objects. It is also apparent that there is a need for interaction techniques such as the *joint control* at the interface of both navigation and manipulation, as handling them separately leads to recurrent issues. Future work will develop these techniques, notably by applying them to devices or proxies with more contact points, to other types of simulations such as deformable objects, or to interfaces with more degrees of freedom. Finally, interaction techniques presented in this chapter were used in simple virtual environments, with a very simple task, and future work should assess their applicability and usefulness for real applications such as medical training or industrial prototyping.

Acknowledgement This research is supported in part by ANR (project Mandarin—ANR-12-CORD-0011).

References

1. Guiard, Y. (1987). Asymmetric division of labor in human skilled bimanual action: the kinematic chain as a model. *Journal of Motor Behavior*, *19*(4), 486–517.
2. Waldron, K., & Tollon, K. (2003). Mechanical characterization of the immersion corp. haptic, bimanual, surgical simulator interface. In B. Siciliano & P. Dario (Eds.), *Springer tracts in advanced robotics: Vol. 5. Experimental robotics VIII* (pp. 106–112). Berlin: Springer.
3. Loi-Wah, S., Van Meer, F., Bailly, Y., & Yeung, C. K. (2007). Design and development of a da Vinci surgical system simulator. In *Proc. of international conference on mechatronics and automation* (pp. 1050–1055).
4. Li, S., Frisoli, A., Avizzano, C. A., Ruffaldi, E., Lugo-Villeda, L. I., & Bergamasco, M. (2009). Bimanual haptic-desktop platform for upper-limb post-stroke rehabilitation: practical trials. In *Proc. of IEEE international conference on robotics and biomimetics* (pp. 480–485).
5. Hulin, T., Sagardia, M., Artigas, J., Schaetzle, S., Kremer, P., & Preusche, C. (2008). Human-scale bimanual haptic interface. In *Proc. of 5th international conference on enactive interfaces* (pp. 28–33).
6. Faeth, A., Oren, M., Sheller, J., Godinez, S., & Harding, C. (2008). Cutting, deforming and painting of 3d meshes in a two handed viso-haptic vr system. In *Proc. of IEEE virtual reality* (pp. 213–216).
7. Dominjon, L., Lécuyer, A., Burkhardt, J.-M., Andrade-Barroso, G., & Richir, S. (2005). The "bubble" technique: interacting with large virtual environments using haptic devices with limited workspace. In *Proc. of the first joint eurohaptics conference and symposium on haptic interfaces for virtual environment and teleoperator systems, WHC'05* (pp. 639–640). Los Alamitos: IEEE Computer Society.
8. Ott, R., De Perrot, V., Thalmann, D., & Vexo, F. (2007). Mhaptic: a haptic manipulation library for generic virtual environments. In *Proc. of international conference on cyberworlds, CW'07* (pp. 338–345). Los Alamitos: IEEE Computer Society.
9. Talvas, A., Marchal, M., Nicolas, C., Cirio, G., Emily, M., & Lécuyer, A. (2012). Novel interactive techniques for bimanual manipulation of 3d objects with two 3dof haptic interfaces. In *Proc. of EuroHaptics (1)* (pp. 552–563).
10. Dominjon, L., Perret, J., & Lécuyer, A. (2007). Novel devices and interaction techniques for human-scale haptics. *The Visual Computer*, *23*(4), 257–266.

11. de Pascale, M., Formaglio, A., & Prattichizzo, D. (2006). A mobile platform for haptic grasping in large environments. *Virtual Reality*, *10*, 11–23.
12. Peer, A., & Buss, M. (2008). A new admittance-type haptic interface for bimanual manipulations. *IEEE/ASME Transactions on Mechatronics*, *13*(4), 416–428.
13. Formaglio, A., Prattichizzo, D., Barbagli, F., & Giannitrapani, A. (2008). Dynamic performance of mobile haptic interfaces. *IEEE Transactions on Robotics*, *24*(3), 559–575.
14. Peer, A., Unterhinninghofen, U., & Buss, M. (2006). Tele-assembly in wide remote environments. In *Proc. of 2nd international workshop on human-centered robotic systems*.
15. Murayama, J., Bougrila, L., Luo, Y., Akahane, K., Hasegawa, S., Hirsbrunner, B., & Sato, M. (2004). Spidar g&g: a two-handed haptic interface for bimanual vr interaction. In *Proc. of EuroHaptics* (pp. 138–146).
16. Garcia-Robledo, P., Ortego, J., Barrio, J., Galiana, I., Ferre, M., & Aracil, R. (2009). Multifinger haptic interface for bimanual manipulation of virtual objects. In *Proc. of IEEE international workshop on haptic audio-visual environments and games* (pp. 30–35).
17. Walairacht, S., Koike, Y., & Sato, M. (2000). String-based haptic interface device for multifingers. In *Proc. of IEEE virtual reality*.
18. Endo, T., Yoshikawa, T., & Kawasaki, H. (2010). Collision avoidance control for a multifingered bimanual haptic interface. In *Proc. of international conference on haptics—generating and perceiving tangible sensations: part II, EuroHaptics'10* (pp. 251–256). Berlin: Springer.
19. Barbagli, F., Salisbury, K., & Devengenzo, R. (2003). Enabling multi-finger, multi-hand virtualized grasping. In *Proc. of IEEE international conference on robotics and automation* (Vol. 1, pp. 809–815).
20. Barbagli, F., Frisoli, A., Salisbury, K., & Bergamasco, M. (2004). Simulating human fingers: a soft finger proxy model and algorithm. In *Proc. of 12th international symposium on haptic interfaces for virtual environment and teleoperator systems* (pp. 9–17).
21. Novint. www.novint.com/index.php.
22. Sensable Technologies. http://www.sensable.com/.
23. Conti, F., Barbagli, F., Balaniuk, R., Halg, M., Lu, C., Morris, D., Sentis, L., Warren, J., Khatib, O., & Salisbury, K. (2003). The CHAI libraries. In *Proc. of eurohaptics* (pp. 496–500).
24. SenseGraphics. http://www.h3dapi.org/.
25. de Pascale, M., & Prattichizzo, D. (2007). The haptik library: a component based architecture for uniform access to haptic devices. *IEEE Robotics & Automation Magazine*, *14*(4), 64–74.
26. Fischer, A., & Vance, J. M. (2003). Phantom haptic device implemented in a projection screen virtual environment. In *Proc. of workshop on virtual environments, EGVE'03* (pp. 225–229). New York: ACM.
27. Isshiki, M., Sezaki, T., Akahane, K., Hashimoto, N., & Sato, M. (2008). A proposal of a clutch mechanism for 6dof haptic devices. In *Proc. of 18th international conference on artificial reality and telexistence* (pp. 57–63).
28. Zhai, S. (1998). User performance in relation to 3d input device design. *SIGGRAPH Computer Graphics*, *32*(4), 50–54.
29. Ullrich, S., Rausch, D., & Kuhlen, T. (2011). Bimanual haptic simulator for medical training: system architecture and performance measurement. In *Joint virtual reality conference of EuroVR—EGVE*.
30. Cutler, L. D., Fröhlich, B., & Hanrahan, P. (1997). Two-handed direct manipulation on the responsive workbench. In *Proc. of symposium on interactive 3D graphics, I3D'97* (pp. 107–114). New York: ACM.
31. García-Robledo, P., Ortego, J., Ferre, M., Barrio, J., & Sánchez-Urán, M. A. (2011). Segmentation of bimanual virtual object manipulation tasks using multifinger haptic interfaces. *IEEE Transactions on Instrumentation and Measurement*, *60*(1), 69–80.
32. Borst, C. W., & Indugula, A. P. (2006). A spring model for whole-hand virtual grasping. *Presence: Teleoperators & Virtual Environments*, *15*(1), 47–61.

33. Pouliquen, M., Duriez, C., Andriot, C., Bernard, A., Chodorge, L., & Gosselin, F. (2005). Real-time finite element finger pinch grasp simulation. In *Proc. of the first joint eurohaptics conference and symposium on haptic interfaces for virtual environment and teleoperator systems* (pp. 323–328).
34. Jacobs, J., & Froehlich, B. (2011). A soft hand model for physically-based manipulation of virtual objects. In *Proc. of IEEE VR* (pp. 11–18).
35. Garre, C., Hernandez, F., Gracia, A., & Otaduy, M. A. (2011). Interactive simulation of a deformable hand for haptic rendering. In *Proc. of IEEE world haptics conference* (pp. 239–244).
36. Jacobs, J., Stengel, M., & Froehlich, B. (2012). A generalized god-object method for plausible finger-based interactions in virtual environments. In *Proc. of IEEE symposium on 3D user interfaces* (pp. 43–51).
37. Colgate, J. E., Stanley, M. C., & Brown, J. M. (1995). Issues in the haptic display of tool use. In *Proc. of international conference on intelligent robots and systems* (Vol. 3, pp. 140–145).
38. Bergamasco, M., Avizzano, C. A., Frisoli, A., Ruffaldi, E., & Marcheschi, S. (2006). Design and validation of a complete haptic system for manipulative tasks. *Advanced Robotics, 20*(3), 367–389.

Chapter 4
Modular Multi-finger Haptic Device: Mechanical Design, Controller and Applications

Ignacio Galiana, Jorge Barrio, Jose M. Breñosa, and Manuel Ferre

Abstract In this chapter, a scalable, multi-finger haptic device based on modular configuration is presented. The mechanical design is based on a modular configuration with a redundant degree of freedom in which each module represents one finger. Mechanical configuration has been optimized to provide a device, which is as transparent as possible to the user. A general description of the control requirements and the implementation to control these types of devices are presented. Applications of modular multi-finger haptic devices include advanced virtual manipulation and simulators for training precise manual tasks. In this chapter, applications for precise manipulation including haptic and visual feedback are presented. Three scenarios have been developed in order to analyze human factors, train manual manipulations and test the performance of the system: (i) a simulator to train physiotherapists to do rehabilitation procedures, (ii) manipulation of fragile objects, and (iii) collaborative manipulation to lift an object between two users.

4.1 Introduction

Haptic devices are mechatronic systems that allow users to interact with virtual or remote environments perceiving object impedance by reflecting forces during manipulation tasks. This kind of devices is typically integrated into multimodal setups that provide haptic, visual, and audio information related to the manipulation performed by the user. Haptic devices are required to read the user's hand- (or finger-) position and display forces that represent interaction with the virtual or real environment. During the past years, several multifinger haptic devices have been developed.

In recent years, haptic interfaces have undergone remarkable developments, including the creation of commercialized equipment [1]. Haptic devices have been used for several applications in different fields such as telerobotics [2, 3], medical surgery [4–6], medical rehabilitation [7], industry [8], training and education [9], and entertainment, among others.

I. Galiana (✉) · J. Barrio · J.M. Breñosa · M. Ferre
Centre for Automation and Robotics UPM-CSIC, Universidad Politécnica de Madrid,
C/ José Gutiérrez Abascal 2, 28006 Madrid, Spain
e-mail: ignacio.galiana@upm.es

I. Galiana, M. Ferre (eds.), *Multi-finger Haptic Interaction*,
Springer Series on Touch and Haptic Systems, DOI 10.1007/978-1-4471-5204-0_4,

In this chapter, the mechanical design and controller of a multi-finger haptic device that is based on a modular design philosophy in which each module is dedicated to a finger will be described. Configurations of 2 and 3 fingers are achieved by combining different modules with a redundant DoF in order to obtain a wide, collision-free workspace. The solution adopted for this device is suitable for dexterous manipulations with up to 3 fingers in which a human wrist workspace is sufficient; in case a wider physical workspace is needed for a grounded haptic device to achieve a particular application, a more complex solution where a haptic device is mounted on a robot arm will be required.

The rest of this chapter is organized as follows. Section 4.2 is focused on describing the mechanical design and evaluation of the MasterFinger. Section 4.3 shows the control architecture developed for this haptic device. Section 4.4 describes several applications including a training simulator for medical rehabilitation, evaluation of haptic interaction when manipulating fragile objects and when performing collaborative manipulations. These applications highlight the performance of this multi-finger haptic device and the benefits of haptic feedback for advanced manipulations. Finally, conclusions about the work described in this chapter are provided.

4.2 Mechanical Design

In order to achieve a higher degree of immersion, haptic devices have to be optimized mechanically to have a proper workspace, to display adequate forces to the user in order to achieve a realistic interaction. The resulting mechanical structures for these devices have to be ergonomic since they are in constant contact with the operator fingers or hand. It will thereby allow comfortable configurations while the user perceives the feedback forces that represent the interaction with virtual or remote objects. In this section, the concept as well as the design philosophy for the mechanical structure of a multi-finger haptic device can be configured up to three fingers due to its modular configuration.

4.2.1 Design Philosophy

A single finger is usually suitable for applications designed for palpation or exploration of objects. However, two fingers are required to perform advanced manipulations such as grasping. The number of fingers involved in the interaction used in a task depends on the morphology of the grasp adopted in order to manipulate a particular object. When performing complex manipulation tasks, tactile information is used to handle objects with precision, so that fingers perceive physical information such as weight, shape, edges, etc. This implies that the more fingers being used, the more information is gathered from the object.

The concept of modularity allows easy scalability of the device to suit the number of fingers required for a particular task or application. Following this modular design methodology, a haptic module applied to a single finger can be designed, and then combined with other modules by adding redundant degrees of freedom (DoF) in order to achieve multi-finger configurations that guarantee a high workspace without collisions. It will improve the user's sense of immersion in remote or virtual environments, increasing the realism and dexterity of the virtual or remote interactions. Based on this goal, a haptic device that can be used to control three-finger robotic hands [10], or to perform advance virtual manipulations will be designed. The mechanical design is optimized to obtain a high workspace, manipulability, inertia and force feedback.

Main objective when designing interfaces is to obtain a mechanical device as 'transparent' as possible. Ideally, the teleoperation system would be completely transparent so that the operator feels as if he/she was directly interacting with the remote or virtual environment [11]. Several transparency measures are defined in the literature, the most common are the following: (i) A system is considered transparent if the master and slave's position and force responses are identical respectively, no matter what the object dynamics are [12]; (ii) A transparent system requires that the impedance transmitted to or "felt" by the operator equals the environmental impedance that the human operator is interacting with [13].

However, to achieve absolute transparency, haptic devices should have no inertia, no friction, or infinite bandwidth. Unfortunately, these features are unattainable and are opposed to each other; this means there will be a trade-off between different characteristics.

Moreover, design complexity increases with the number of fingers of the resulting device, mainly due to the difficulty in obtaining a sufficiently large and collision-free workspace. The requirements considered for the mechanical design of this multi-finger haptic device are as follows:

1. Structures should be large enough to achieve workspace that is wide enough to perform tasks in a natural way with multiple fingers.
2. The apparent inertia of the interface should be as low as possible. This is achieved by allocating motors close to the base and reducing the weight of the mechanical structure and end-effector [14].
3. The global stiffness of the structure must be sufficient to prevent excessive deviations at the end effector.
4. At least 3 DoF are required to actuate the kinematic chain from the base of each finger in order to be able to exert forces on the tip of each finger in any direction. Moreover, in order to achieve a realistic interaction at least 3 N are required in any direction within the workspace [15].
5. Scalability principle: a basic finger module should be easy to integrate with more modules to achieve a multi-finger haptic device. To make this possible, the structure of the basic module must be as simple and compact as possible.

As a result in order to have a proper design of a modular multi-finger haptic device, the following requirements have to be taken into account: high workspace, low inertia, high stiffness, ability to exert large forces in any direction within the workspace and scalability. A modular haptic device has been designed based on

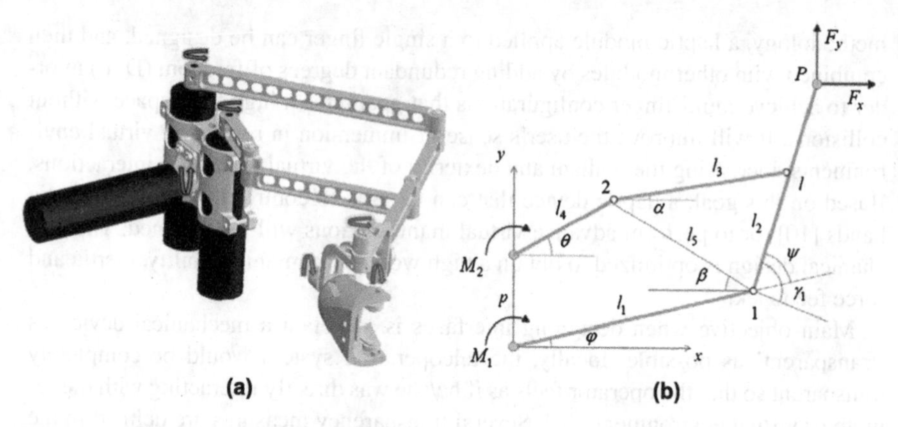

Fig. 4.1 (**a**) Haptic Modúle for one finger, with 3 actuated DoF at motors, another three with free orientation. (**b**) Five-bar Mechanism configuration including 2 actuators

the optimization of these design principles; following subsections describe the basic module design and describes configurations to achieve 2 and 3 fingers interaction.

4.2.2 One Finger Module

As each finger interacts through a single module in a scalable haptic device, a finger can be considered as the basic module; hence, each module has to be fully functional for one finger. The structure of the module should be simple and compact in order to be easily scalable, and thereby avoiding collisions between modules in multi-finger configurations.

The structure for the kinematic chain of each module consists of a series-parallel structure with a total of 6 DoFs, as shown in Fig. 4.1a. The first three DoFs are actuated in order to provide forces in any direction to the finger. The last three DoFs are passive and allow any orientation to the end effector.

The parallel structure, Fig. 4.1b, is designed as a five-bar mechanism with 2 actuated joins, which is connected to the first actuator in a serial configuration.

The 5-bar mechanism can be configured with two different configurations: "elbow-out" and "elbow-in". It has been demonstrated that the "elbow-out" configuration provides a better performance within the workspace [16].Both, material and shape of the structure have also been taken into account in this study, aluminum alloy provides high stiffness and light-weight to the structure.

The mechanical structure is optimized with respect to performance measures such as kinematic device workspace, payload, isotropy, inertia, and friction in joints.

In order to reduce device inertia, all actuators have been located near the base. Three DC motors from Maxon Motors with a 14:1 gearbox have been used for this basic module. Lengths of links affect the workspace amplitude and performance of the 5-bar mechanism [17, 18]. Hence, they have to be optimized in order to achieve

Table 4.1 Main mechanical specifications of single finger module

Property	Value
Minimum of maximum exertable forces for Nominal position	3 N
Inertia in Nominal position (apparent mass at tip)	
Without encoder gimbal	52 g
With encoder gimbal	128 g
Stiffness in Nominal position	1.54 N/mm
Weight of the device	650 g
Bandwidth	8 Hz

a high dexterous workspace with sufficient load capabilities. These conditions have been implemented in a multi-criteria design optimization in order to optimize the mechanical design dimensions [16]. Multi-criteria design is applied so as to obtain links lengths that optimize the Global Condition Index (GCI) and performance indexes [16].

The nominal torque for the actuators was selected in order to obtain 3 N of force in every direction within the workspace at the fingertip contact point.

Finally, the obtained specifications of the single finger module are given in Table 4.1. In the plane of the five-bar mechanism, the apparent mass for the middle position of the workspace (20 cm from the motor axis) is only 39 g, 30 % of this is due to the rotor inertia. The apparent mass about the axis of the engine in serial position is 52 g. Additionally, 76 g weight of the thimble must be added to the apparent inertia. The mechanical stiffness is 1.54 N/mm and was estimated using Finite Element Analysis (FEA) with SolidWorks simulation software, where the stiffness of the device is calculated by simulating an external force at the end-effector and measuring the displacement obtained.

The bandwidth of the proposed system has been measured by commanding a sinusoidal trajectory and acquiring the encoder position of the motors. These measurements were taken at the plane of the five-bar mechanism in the middle position of the workspace (20 cm from the motor axis). The resulting frequency (for −3 dB) is approximately 8 Hz. The current bandwidth is limited to 8 Hz mainly due to the selected actuator planetary gear.

4.2.3 Two-Finger Haptic Device

Based on the basic module previously described, the system can be easily scaled by combining modules. This section describes the mechanical design of a two-finger modular haptic device that has been made by combining two basic haptic modules in opposite position and a redundant DoF in the base. In doing so, it avoids collisions between different modules to allow a wide, ergonomic workspace.

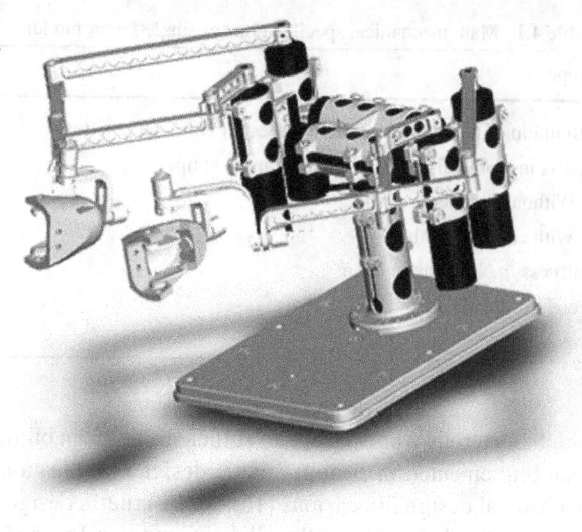

Fig. 4.2 CAD design for the two-fingered haptic device for advanced manipulation

The two-finger haptic device consists of a mechanical structure that allows the manipulation of objects with the index and thumb fingers. This device, as shown in Fig. 4.2, is suitable for bimanual manipulation tasks [19] giving the user a wide collision-free workspace.

The intersection workspace for the two modules can be estimated by an ellipse of 21 × 16 cm axes. By adding a redundant DoF that connects both modules to the base, this working area is significantly increased. Figure 4.2. shows the designed 2-finger haptic device; where all its actuators are located near the base of the structure in order to reduce the inertia. Moreover, the redundant motor is located in the normal axis going through the center of mass of the modules' common base. In such manner, a broad operative workspace that allows wrist movements by avoiding collisions between the links when there is a change in the orientation of the hand is achieved. This device is optimized to grasp objects with thumb and forefinger allowing the user to interact with virtual environments and perform tasks in an easy and comfortable way.

This two-fingered haptic device has been used in various applications for manipulating virtual objects. Figure 4.3 provides an example of bi-manual virtual manipulation where the haptic devices are placed in an inverted position. In this case, a virtual box is grasped and lifted by either one or two hands [20].

4.2.4 Three-Finger Haptic Device

Considering the concept of scalability as in previous section, a new module was added to the two-fingered device to obtain a three-fingered haptic device. This device also has a redundant axis at the common base of the modules, to ensure an adequate workspace by avoiding collisions among different links.

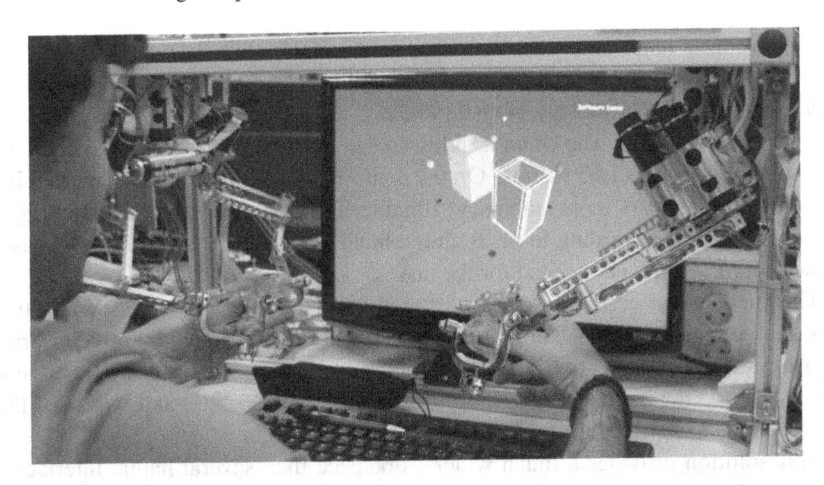

Fig. 4.3 Two-fingered haptic devices performing a bimanual task. Index and thumb are used for manipulating virtual objects

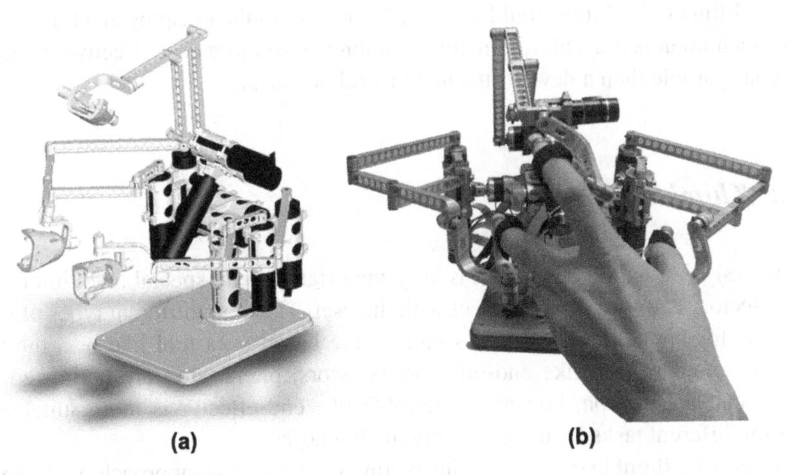

Fig. 4.4 CAD design for the three-fingered haptic device for advanced manipulation

In order to obtain a suitable workspace, the third module is placed in an intermediate position with respect to the other two modules so as not to limit the global reachable area of the hand. An orientation of 45° minimizes the number of collisions among the rest of the modules. As it has been described before, a redundant DoF is actuated from the base to prevent collisions in some particular positions and orientations of the hand and fingers. This redundancy allows free movement of the fingers, hand and wrist, creating a toroidal shape workspace for the hand.

The final configuration is adopted as shown in Fig. 4.4, allowing a wide workspace to the user, which may perform all hand movements including all fingers, wrist and even amplitude movements concerning the elbow joint.

This 3-finger haptic device allows the user to manipulate the virtual or remote environment using the thumb, index and middle finger, or ring finger instead of middle, depending on the type of task to be performed and the gesture required for the manipulation [21]: The thumb, index and middle fingers are used to perform precise manipulations like writing. On the contrary, the configuration of thumb, index and ring fingers is more stable and suitable for heavy loads. The middle finger or ring is selected according to the required hand gesture [22, 23] and on the basis of the accuracy or power needed for the task.

This mechanical design described for modular haptic devices of 1, 2 and 3 fingers, may not be suitable when it is extended to four or five fingers. Problems arising from the high concentration of actuators, make it difficult to locate them near the base of the device without a significant increase in the mass and inertia of the device, as well as collisions among different modules occur.

This solution provides a much wider workspace than several haptic interfaces, while similar in complexity. By contrast, some of these existing multi-finger haptic devices offer larger workspace, but at the cost of greater complexity and in need of a robotic arm to transport the haptic device [24].

The 3-fingered solution could be suitable for most of the grasping and functionalities of a human hand. This alternative solution is also more cost-effective, portable and configurable than a device attached to a robotic arm.

4.2.5 Thimble

In the design of haptic devices, it is very important to pay special attention to the end-effector, as it is in direct contact with the user. There are different types of end-effectors for haptic devices; most solutions are based on a tool like or a thimble like end-effector. Tool like end-effectors (scissors, pliers, etc.) are optimized for a particular application, however, a thimble like end-effector is more suitable to perform different tasks by using directly the fingertips.

In case of a thimble design, it is interesting to consider an approach based on an adaptable thimble for any finger size. Because of this, there are many requirements to consider in a thimble for a multi-finger haptic device:

1. Adaptability to different fingers of the human hand, having a structure as small as possible for avoiding accidental collisions.
2. Ergonomics: the user feels comfortable when using the haptic device.
3. The clamping force of the user's finger into the thimble should be just right to ensure that the finger does not slip, and quite low enough so that the strength affects the user's tactile perception as less as possible.
4. It must be as light as possible to avoid inertia at the end of the haptic device kinematic chain, since the inertia directly affects the user's perception.
5. The thimble must join the haptic device so that only forces are exerted, without undesired torques to the user.

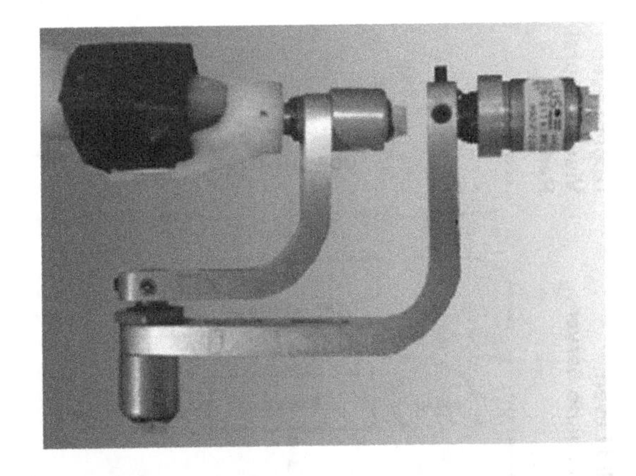

Fig. 4.5 Thimble with gimbal configuration. Gimbal has 3 rotational DoF that allows all orientation for the thimble

The thimble has been ergonomically designed to allow exploration of objects and to achieve a realistic grasping feeling. It has been designed to suit different sizes of fingers by using an elastic band that ensures contact while avoiding slips as shown in Fig. 4.5.

The thimble design meets the requirements described previously evolving the human finger geometry and keeping it in constant contact to perceive objects through the sense of touch.

The thimble is attached to the 5-bar mechanism through a configuration in gimble allowing any orientation in the workspace. These rotations are measured from the encoders attached to each axis of rotation. The intersection of the three axes is located at the fingertip to avoid torques, so only forces are reflected on the operator.

4.3 Control Architecture for a Multi-finger Haptic Device

In this section, the different components and requirements of the control architecture for the modular multi-finger haptic device presented in previous section will be described. The developed system provides haptic, visual and aural feedback to the user.

In order to reach smooth transitions in graphics, at least 50 Hz update frequency rate is needed for monocular visual information. If stereoscopic vision is required, this frequency should be increased to at least 100 Hz. Force feedback has the higher requirements for the control system not only related to provide enough realism but also to assure stability, 1 KHz update rate is necessary to achieve the required haptic fidelity [25].

Hence, to satisfy these requirements, a real-time control architecture that processes and controls the required signals in real-time has been developed. This control architecture of virtual manipulation is made up of three software layers: the low-level controller, the scenario server and the visual and aural renderer. In this section, each of the layers shown in Fig. 4.6 are described.

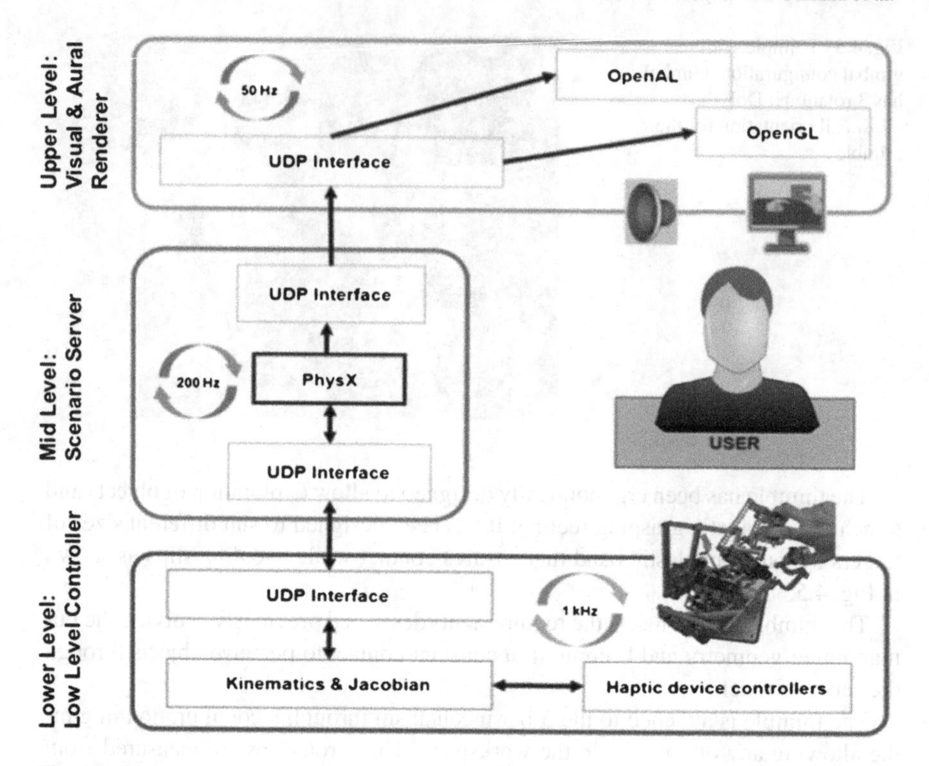

Fig. 4.6 Software architecture developed to control a Multi-finger haptic device

4.3.1 Low-Level Controller

This layer is in charge of processing the measurements from the sensors, calculating the position and orientation of the end-effectors for each independent module, controlling each motor current and torque, and communicating via Ethernet with the immediately upper layer (Scenario Server). Furthermore, this layer synchronizes all the other processes and computes mathematically complex kinematic and Jacobian calculations. Once the position and orientation of the different thimbles are calculated, they are sent to the upper layer, which is the physical engine. In return, the forces to be reflected are received from the physics engine, and then the low-level controller calculates the torques to be produced in each motor by computing the Jacobian matrix.

In order to avoid unwished vibrations to the user and provide realistic forces, this software level must run at 1 kHz [25]. Delays in this layer can lead to inconsistent situations where real fingers are in very different positions from the virtual ones. This can make the haptic interface reflect undesired forces, vibrations or even become unstable if not considered in the controller design process. Hence, a hard real time system is required to avoid these effects

To better understand the chosen control architecture, all the signals have to be processed by the controller. Each finger module consists of a five-bar linkage mech-

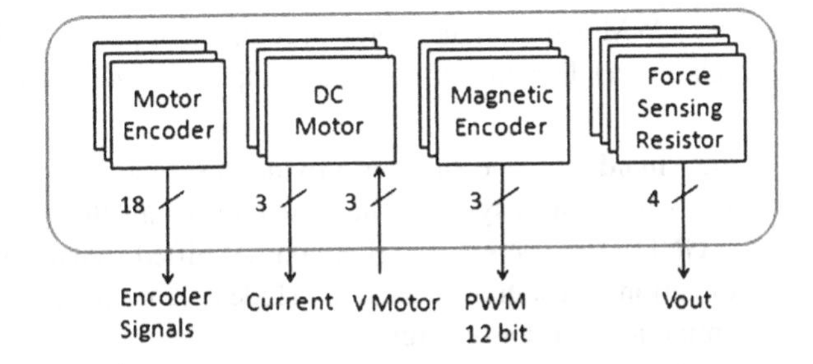

Fig. 4.7 Control Signals for one module of the 3-finger Haptic Device

anism with three actuated degrees of freedom (DoF) and three non-actuated DoF at the end-effector. First three DOF allow exerting forces in any direction within the workspace, while the 3 DoF at the end effector allow a free orientation of the finger tip with no torque transmission thanks to a gimble configuration as shown in Fig. 4.1 [26, 27].

In addition, magnetic encoders are placed in the gimble to determine the orientation of the fingers and 4 contact force sensors are placed in the thimble to estimate the forces perceived by the user when interacting with the virtual environment [28–30]. To sum up, every module has to process and/or control the following signals:

- 3 motors each with each its PWM signal, current measurement and encoder.
- 3 magnetic sensors at the gimble.
- 4 force sensing resistors at the thimble.

As shown in Fig. 4.7, a total of 28 signals for sensor information and 3 control signals are required to manage each finger. As three fingers and a redundant DoF make up this device, a total of 91 signals for sensor information and 10 control signals have to be managed in real time by the control system. The signals of the 3-finger haptic device will be processed in this layer.

As the system interacts with a virtual collaborative environment, a distributed architecture is required. In this architecture, each module is connected using an Ethernet connection with an IP/UDP protocol when in a local network or the IP/BTP [31] protocol when in a nonlocal network. This connection will satisfy the frequency required by MasterFinger-3 and should also be easy to install, maintain and update.

The equipment used must be powerful enough to process all the required signals and to close the control loop with a frequency that guarantees a good performance of the system (1 kHz) [25].

It also has to be considered that although electronics could be specifically designed for one module and managed with a centralized control, it is preferable to use a single computer where everything can be integrated and establish connections among different degrees of freedom via software.

PWM To actuate the DC motors, a PWM is calculated using the duty cycle provided by the PowerPC and a 10-MHz clock signal, as torque in each motor is estimated from current measurement, it is also needed to read this to close the current loop.

This PWM is sent to the motors through their power wires (positive and negative in each motor). For this purpose, two output power connections are needed.

Motor Encoder The encoder provides a simple square signal that is further processed by the control system. There are three channels (A, B, I) and two power wires (Vcc, GND). Channels A and B are phase shifted signals that are used to determine the direction of rotation. Channel I (Index) is used as reference point for precise determination of rotation angle.

The line driver produces complementary signals of each channel, so a total of eight wires are needed to manage encoders: two power outputs and six signal inputs.

Gimble Encoders The orientation of the fingertip is obtained by magnetic encoders located at the gimble's rotational axes. The encoders used for this system are MA3-P12 from US Digital. The MA3 is a miniature rotary absolute magnetic shaft encoder that reports the shaft position over 360° with no stops or gaps.

MA3-P12 produces a 12-bit PWM output, so it adds two output power connections to the system requirements and one signal input for the PWM angular position.

Force Sensors The contact sensors located inside the thimbles provide information about the force perceived by the user.

Four Tekscan's FlexiForce A201 are used in each thimble, one of them located under the finger, one at the end of the finger, and one on either side of the finger. The FlexiForce sensing area is treated as a single contact point to estimate the normal and tangential forces perceived by the user [28, 29].

This resistive sensor needs one voltage input and a voltage output for measuring the force; therefore a power output and one input signal are needed.

The following sections will address the hardware chosen for the control equipment and the software architectures designed for that purpose.

Electronic Hardware Due to the high system requirements for multi-finger haptic applications and to the high number of signals (81) that have to be managed and controlled, selecting the correct hardware is essential to guarantee a high performance.

For this three-fingered haptic device, a PXI chassis system from National Instruments (Fig. 4.8a) [32] was chosen. This system consists of three parts: A Real Time Controller, an FPGA and the power electronics.

Both the FPGA and the Real Time Controller can be programmed in LabView graphical development environment thus reducing the overall programming time for a non-expert in VHDL hardware definition language and C/C++ programming.

While the user is interacting with the virtual scenario by means of the described multi-finger interface, the Scenario Server calculates the resulting interaction force that has to be exerted to the user. In this section, the different components of the electronic are explained.

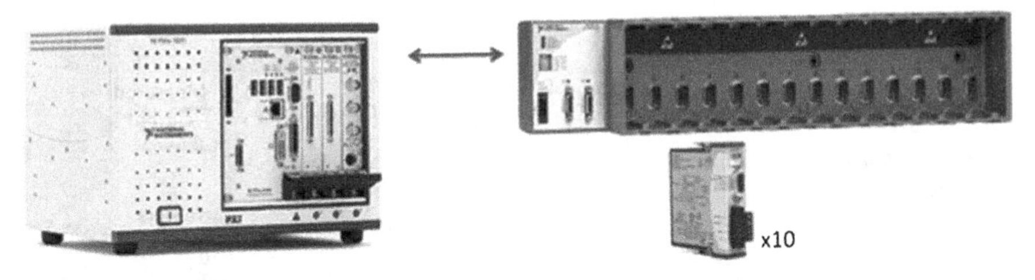

Fig. 4.8 (**a**) PXI System with RT Controller, (**b**) FPGA Virtex-5 with Chassis to connect different modules. (**c**) Full H-Bridge Brushed DC Servo Drive Module

Real Time Controller The PXI chassis incorporates a 1.9 GHz Dual-Core Real-Time embedded controller NI PXIe-8102 running VxWorks. This Interaction force calculated by the Scenario Server is sent to the Real-Time Controller by the UDP interface.

The controller is used for programming the complex mathematic calculations (Jacobian, Kinematic Equations, Controller Equations, etc.) to transform the interaction force to a current setpoint in the actuators and to calculate the position (x, y, z) from the encoders position of the actuators.

The RT Controller communicates with the FPGA via a MXI high-speed cable connection to receive the encoders' position and to send the current set point.

FPGA The PXI Chassis communicates via a high-speed MXI connection with a FPGA Virtex-5 integrated in a compact module that allows connecting up to 14 modules with different functionalities directly to the FPGA (Acquisition of Analog and Digital Signals, power electronics, etc.) (Fig. 4.8b).

The FPGA Virtex-5 is configured using LabView graphic programming language. For this application, the FPGA is used to:

1. Acquire and process Signals: encoders to calculate position and velocity of the device and current measurement.
2. Actuators' Current Loop: A PI controller is implemented to control the current of the actuators and generate a PWM in each actuator to exert to the user the required force.
3. Communication with the Real Time Controller: The encoder position of the actuators is sent to the RealTime Controller and the Current Command required to exert the interaction force to the user is received from the RT Controller.

Power Electronics For each actuator a Full H-Bridge Brushed DC Servo Drive Module (Fig. 4.8c) from NI directly connected to the FPGA was used. Each of these modules provides a measurement of the current that circulates over the DC actuator, has a data acquisition for the Encoder and provides power to the DC actuators used.

The low level controller runs in the PXI platform that, as described before and consists of a Real-time controller that communicates with the other modules via Ethernet and a FPGA to control the system and acquire the different signals.

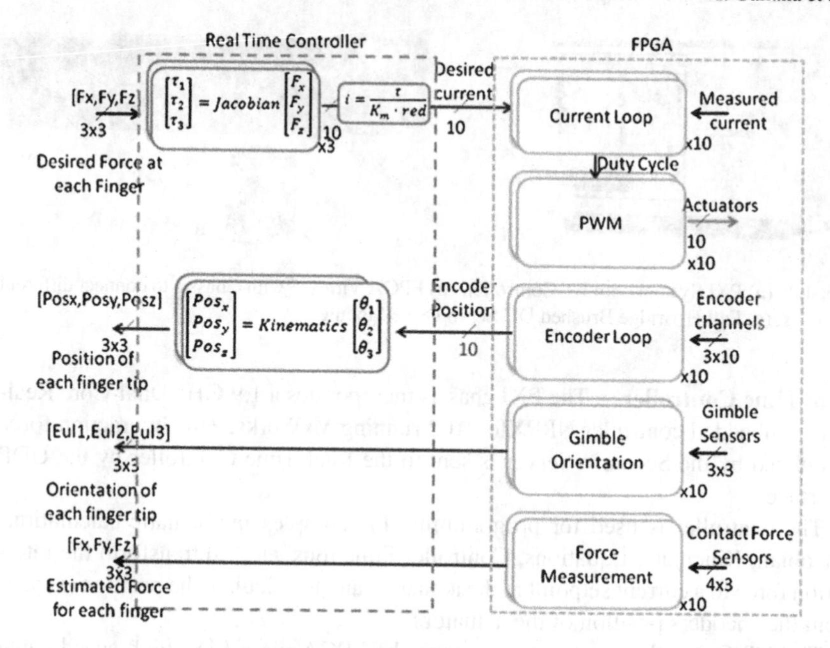

Fig. 4.9 Low-Level Control Architecture. All signals processed by each module are detailed

The Low-level controller has been programmed using LabView graphical development environment (LabView RealTime module for the RT Controller and LabView FPGA module for the Virtex- 5 FPGA), a detailed description of the tasks that this controller does is shown in Fig. 4.9.

4.3.2 *Layer 2: Scenario Server*

The software architecture design is a key point for developing haptic applications. Different options such as the union between haptic and visual interfaces, the software election or how the virtual scenarios are recreated, should be analyzed. Moreover, the software architecture should be as independent as possible from the hardware layer.

High-level software is responsible for updating the scenario, especially for calculating force feedback and processing visual and aural information. The suggested solution shown in Fig. 4.6, splits the software into different layers that run in the different electronic devices as described in the previous section and communicate via an Ethernet connection.

As it has been explained before, the software solutions proposed for low level layers have to be developed ad hoc, as they are very dependent on the device's hardware and it was a development by the Centre for Automation and Robotics UPM-CSIC and have been explained in previous layer. Meanwhile, a wide range of commercially available software products can be used for high-level layers, as they are performed in an abstract level.

In this module, very heavy calculations are processed. In this place, the virtual scenario is created and updated with the data received from the low-level controller. Collision detection, interaction calculations and body dynamics are some of the tasks assigned to this layer. It also has to communicate the scenario status to the graphics interface and the interaction forces to be exerted to the virtual fingers to layer 1.

The physical engine is in charge of: identifying collisions, estimating external forces resulting from the collisions (including frictional forces) and to produce a correct response to the residual forces [33].

Due to the high requirements of the physical engine, implementing ad hoc solutions for this layer is very time-consuming and requires big development efforts. To tackle this problem, the adopted solution was to study the software kits commercially available and try to take advantage of their features. In literature, haptic interfaces using commercial physics engines are not very common, most of the available simulators are optimized for a set of commercial devices. This happens mainly because the existing physical engines are often focused on a small set of physical laws and they are usually implemented aiming very accurate but computationally demanding behaviors that cannot be executed in real time.

Some others physical engines can be found working in medical haptic simulators [34]. Some other engines provide a framework where new physical laws can be implemented and included by using software containers.

Related to videogames, many physical engines have been developed recently. They are not optimized for haptics, but they usually have features that are also interesting to this field. It is very common to find fluids, complex rigid bodies and even soft bodies in video games, which gives an idea about the potential of these tools applied in the field of haptic simulation. Some of these engines were tested: ODE, Bullet (which also have the advantage of being open source), Havok and PhysX. Finally PhysX from Nvidia was used because it automatically allocates most of the calculations in the GPU if using Nvidia products which leads to faster, more realistic simulations that fulfill the requirements needed.

4.3.3 Layer 3: Graphic and Aural Renderer

This layer's functionality consists of drawing the virtual objects and fingers and reproducing sounds. As it has been said before, it receives the data from the scenario server.

There are several possibilities regarding software for programming the visual interface, two graphic API's were taken into account for this system: OpenGL and DirectX. DirectX is an API developed by Microsoft and full features (as hardware acceleration) are only obtained when running on Windows based OS. However, OpenGL is an open source API that works under nearly any OS. Its features are sufficient to meet our demands and it also offers high compatibility, which is quite convenient for this project. For the same reasons, OpenAL was chosen for the aural interface.

4.4 Experiments and Applications

In this section several applications and experiments are described in order to show the performance of MF-2 and MF-3 devices and their potential capabilities in different domains, from training precise manual skills for physiotherapists to performing rehabilitation, to perception studies by manipulating fragile objects, or collaborating with someone to lift an object.

4.4.1 Capture of Manipulation Patterns for Simulation of Medical Rehabilitation Procedures

This section describes the development of a model of soft tissue and metacarpophalangeal (MCP) joint behaviour using force measurements taken while performing massage and force treatment techniques. This application shows the potential use of the MF-2 device to record forces applied by therapists when performing a particular technique; once the forces are recorded properties such as joint stiffness in different axes can be identified to create a virtual model of a human body joint with which students can practice interacting through the haptic device. More information about this experiment can be found in [7].

4.4.1.1 Virtual Scenario

To provide reference data, measurements were taken during two techniques that are used in both assessment and treatment: (1) massage of the web between the thumb and index finger, and (2) passive movements of the MCP joint of the index finger. The operator's thumb and index finger were fitted to the thimbles of the MF-2 and the movements were performed with the actuators inactivated. In this way force data could be obtained from the FSRs located in the thimbles, position data from the encoders located in the actuators, and orientation data from the encoders on the axes of the gimbals.

The clinician recorded forces that were applied in performing the techniques and providing descriptions and images. After the recording phase, the other authors sent videos and force traces which were taken while performing the techniques for feedback. This process is reasonable as currently professors are using video and separate force data [35] in their teaching and evaluating students who are in campus or studying from home.

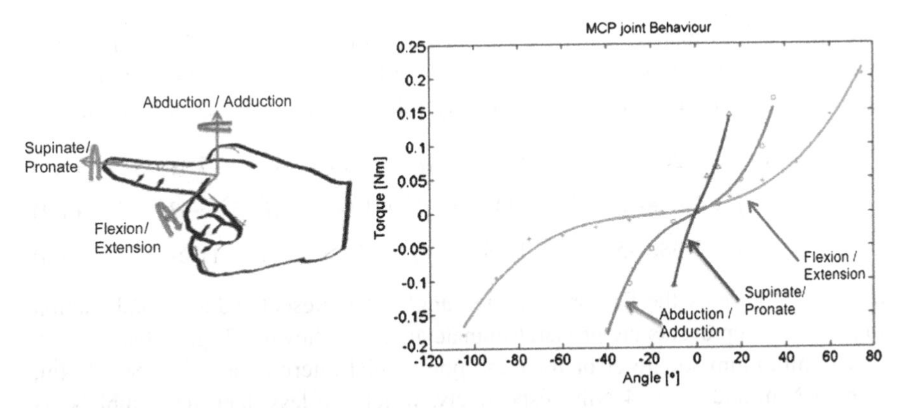

Fig. 4.10 Data obtained for the metacarpal joint of the index finger. This data concerns the three possible rotations in the MCP joint (Reprinted from [7])

The reference data for stiffness characteristics of rotation around each axis of the MCP joint were calculated from the experimental data and integrated into a virtual model of the hand that can be interrogated using the MF-2 haptic device. In the virtual model, the metacarpal bone will be fixed to the ground so the two-finger interface of the MF-2 can be used to manipulate the proximal phalanx of the index finger.

4.4.1.2 Methods for Virtual Model

Joint stiffness is likely to vary with rate of movement so the joint was moved at approximately 3 Hz through a range (12–16 N) when collecting the reference data as this is near the middle of suggested rates of mobilization used in treatment [36]. Force angle plots for each of the three rotation axes were produced from the force and location data measured during performance of the techniques and are shown in Fig. 4.10b.

A minimal square interpolation was fitted to the points for each axis of movement to estimate the joint's behavior. The curves demonstrate that the angular limits of the MCP joint are different in the two directions for each axis.

Because a linear interpolation would result in a non-realistic model, a third order interpolation was used to fit the data. Equations obtained are:

The sensorized thimble, which the modular device has at the end-effector, can be used for creating an approximate model of the characteristics of human joints that are inherently multidimensional and non-linear [37]. The system developed characterizes the stiffness of the metacarpo-phalangeal joint, which is located at the index finger in the axes of rotation. Note that even though the finger can mainly move in the flexion/extension and abduction/adduction angles of rotation, for rehabilitation of the finger, the pronate/supinate degrees of freedom (DoF) should also be considered. The finger was mobilized in the whole range of movements and the force was saved for different angles of the finger; a minimal square polynomial interpolation was calculated to approximate the joint's behavior in every rotational DoF, as shown in Fig. 4.10b. Equations (4.1)–(4.3) show the relationship between the angles and torques (force was applied at approximately 1.2 cm from the center of rotation):

$$\tau(\alpha) = 2.2\text{e}{-7} \cdot \alpha^3 + 1.24\text{e}{-05} \cdot \alpha^2 + 5.56\text{e}{-04} \cdot \alpha + 7.1\text{e}{-3} \quad (4.1)$$

$$\tau(\beta) = 2.04\text{e}{-6} \cdot \beta^3 + 1.44\text{e}{-5} \cdot \beta^2 + 1.6\text{e}{-3} \cdot \beta - 5.9\text{e}{-3} \quad (4.2)$$

$$\tau(\gamma) = 1.68\text{e}{-5} \cdot \gamma^3 - -1.68\text{e}{-4} \cdot \gamma^2 + 7.8\text{e}{-3} \cdot \gamma + 1.1\text{e}{-3} \quad (4.3)$$

where α represents the flexion/extension angle, β represents adduction/abduction angle, and γ represents the pronate/supinate angle, as shown in Fig. 4.10a.

The minimum square error for these polynomial interpolations is 5.8e−3 Nm, 4.6e−3 Nm, and 8.6e−4 Nm, respectively, which are less than the thimble's error. These results are similar to those of previous found in the vitro study of flexion/extension stiffness using freshly frozen fingers from a cadaver [38]. This model was implemented in a simulator and potentially can be used by students to learn and practice rehabilitation procedures.

4.4.1.3 Experiment Study

Two applications are performed using the haptic setup to simulate an assessment and treatment procedure. The assessment consists of an exploration to determine both the quality and quantity of joint movement. The treatment simulation imitates a treatment intended to restore joint function. The maximum force used in producing the reference data was approximately 16 N and the maximum force that can be produced by the MF-2 is 4 N so the reflective forces produced were scaled down by a factor of 1:4.

The joint movement exploration consisted of checking the MCP joint range and rigidity as would be used in a patient to detect unusual boundaries of the movement. In this exploration the MF-2 directly controls the proximal phalanx of the patient's index finger. As shown in Fig. 4.11a. The operator starts moving the joint and can both feel and see the resulting movement on the display. The application records the angles that are achieved in each direction. This exploration simulates the operator identifying boundaries and establishing a baseline to evaluate the effectiveness of treatment. After the session, the operator can repeat the exploration to reassess the boundaries. Future developments can include changes in the force reflection characteristics over time such as improvement following treatment and transient changes from muscle contraction or protective spasm.

Force treatment simulation consists of exerting an oscillatory force with the thumb and the index finger near the boundaries of the MCP joint motion of the virtual hand in order to simulate the forces needed for the treatment. PhysX is only capable of modeling linear stiffness, however as seen in the previous section, the forces required depend of a nonlinear formula. The forces were calculated by other pro-

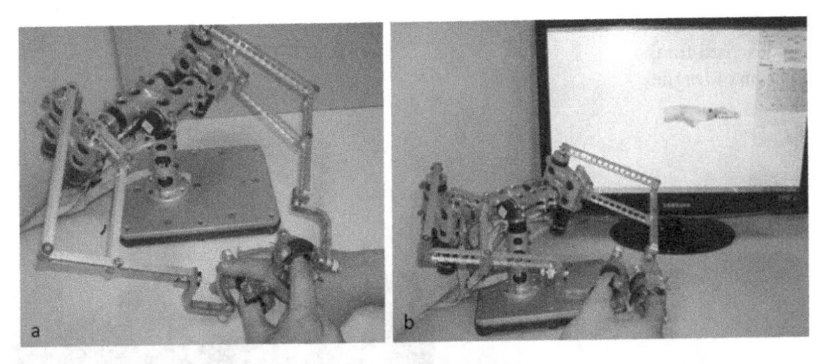

Fig. 4.11 (**a**) Capture of force and position data when mobilizing the MCP joint. (**b**) User interacting with the model obtained from the recordings (Reprinted from [7])

grams based on (4.1)–(4.3) obtained in the previous section. The calculated forces were sent to the MF-2 in order to exert the correct force to the user and to PhysX to update the visual information of the virtual model. Moreover, forces are directly related to the angle of the MCP joint and the penetration of the sphere in the virtual index finger. The developed setup for interacting with the virtual model of the rehabilitation task, is shown in Fig. 4.11b.

4.4.2 Manipulation of Fragile Objects

Precision grasping can only be achieved by means of multiple fingers interacting upon one object. This kind of grasping is commonly used for dexterous tasks, such as grasping light and fragile objects.

Although there are several experiments for evaluating haptic interfaces [39], they normally focus on devices that interact with the objects by means of only one contact point.

This section describes the experiment carried out to evaluate a multi-finger haptic interface, in the manipulation of fragile objects. The main goal of the experiment is to verify whether the haptic interface is suitable for manipulating fragile objects such as an egg, thereby demonstrating that the user can feel a light object and is able to control the force applied to ensure the egg does not break.

The experiment consists of lifting an egg from the floor and placing it back where it was without damaging for a given number of times. Specifically, the same experiment is executed three times under different conditions: first, in the manipulation of a real egg, then, in the manipulation of a virtual object without force feedback and, finally, in the manipulation of a virtual object with force feedback. The order of execution for each part of the experiment is different and randomized for every participant. The results of each execution are registered and processed in order to compare the different performance measures.

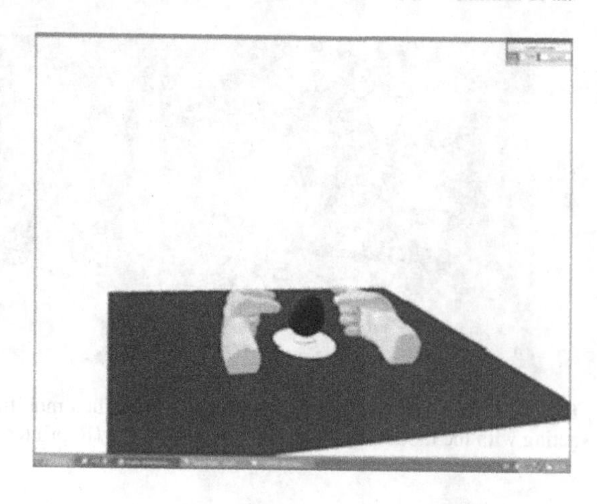

Fig. 4.12 Virtual egg scenario designed for the manipulation of fragile objects

4.4.2.1 Methods for This Experiment

A virtual scenario was developed for this experiment. The scenario contains a virtual model of a real egg that can be moved through the scenario as shown in Fig. 4.12. The implemented model breaks if it falls from a height of over 17 cm or if the user applies a force of more than 14 N when he or she grasps it. The scenario implements a simplified friction model when the user is grasping the egg. If the user does not apply enough force, the egg might slip from the user's fingers and fall on the floor.

The experiment consists of three different tests: real object manipulation (R), virtual object manipulation without force feedback (VNH) and virtual object manipulation with force feedback (VH).

In each test the user must grasp the fragile object (real or virtual) and lift it to a height that is dependent upon the haptic interface's workspace. Then the user must place the object in the original position without breaking it. The object may break if: (a) the user exerts too much force, (b) the force exerted is not enough and the object slips or (c) if the user lets the object fall from above a certain height.

The user must wear the haptic device while handling the real object, to ensure that the user has a similar sensation as when grasping the virtual object. Each user must carry out each experiment sixty times so that a total of 180 lifts will be performed per participant.

Order of tests is randomized for every person. Since there are three tests, there are six possibilities for arranging the tests. The number of people must be a multiple of six in order to obtain the same number of results for each combination.

4.4.2.2 Experiment Procedure

Thirty UPM students volunteered to participate in the experiment. As described in the previous section, they each carried out the test three times (R, VNH and VH).

The system stores the time needed to complete each lift in milliseconds. In order to carry out the experiment, the MasterFinger-2 was mounted upside down on a

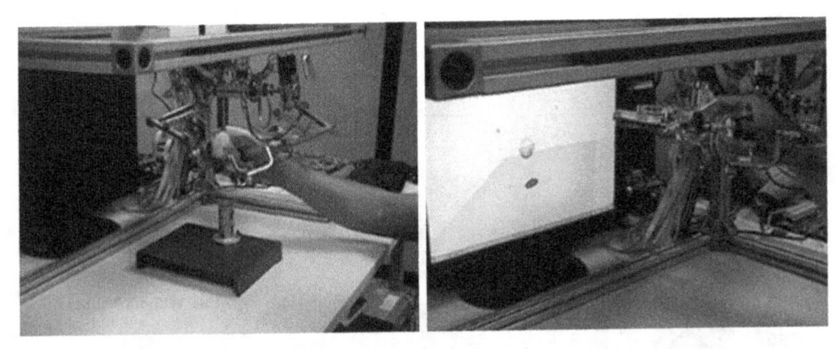

Fig. 4.13 (**a**) Lifting a real egg with the MasterFinger haptic device. (**b**) Lifting a virtual egg with the MasterFinger haptic device

metallic structure. The MasterFinger workspace is more suitable for this experiment since the hand position is more comfortable in this suspended position.

Additionally, behind the haptic device a screen is placed so that the user can see the virtual environment. This setup allows the user to see the haptic interface and the virtual environment at the same time. Figure 4.13 shows the entire setup for the manipulation of a real and a virtual object.

4.4.2.3 Experiments Results

Several results can be obtained that are relevant to evaluating haptic device performance.

First of all, the learning time needed in each test can be measured and compared. Secondly, the effectiveness of real and virtual object handling tasks can be evaluated. Thirdly, the time spent lifting the object in each situation can be compared. An analysis of variance (ANOVA [40]) is used for making comparisons between the different tests (R, VNH and VH). The tests were performed at a significance level of one percent.

Success Rate of the Task Success rate is measured by counting the number of objects that break while lifted.

The results (depicted in Fig. 4.14) clearly show that lifting a fragile object without haptic information is different from lifting a real object. In fact, fifty percent of the eggs lifted without force feedback ended up breaking and only forty percent were correctly manipulated. The main reason for this is that the users are unable to ascertain, from the image alone, whether they are penetrating the object, and therefore the object breaks.

When comparing virtual lifts with force feedback to real lifts, there is no significant difference. In the case of real eggs, 88.16 % are handled correctly, whereas 84.5 %, in the case of virtual eggs with haptics.

An ANOVA test was performed to compare both experiments. Figure 4.15 shows the results of the analysis, wherein lifting a real egg and a virtual egg with force feedback has the same effectiveness.

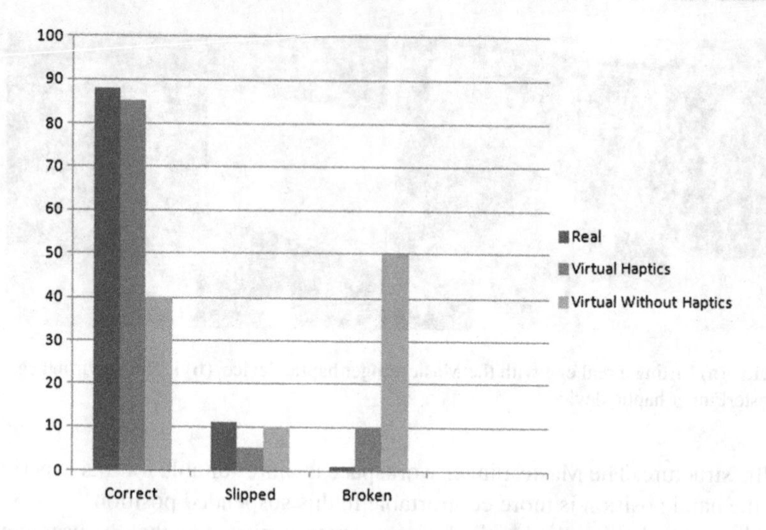

Fig. 4.14 Percentage of successful completions, broken objects and slipped objects when manipulating fragile objects. Comparison of success rates between real manipulation, virtual with haptics and virtual without haptics

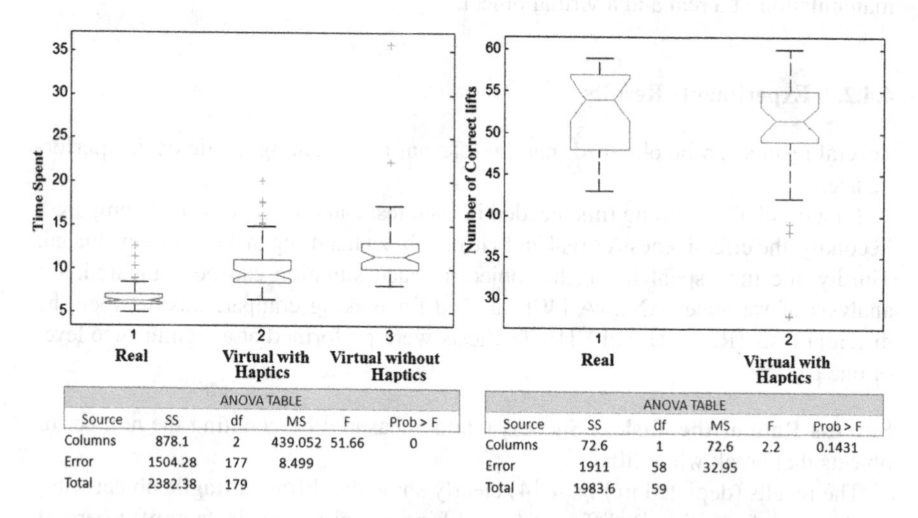

ANOVA TABLE

Source	SS	df	MS	F	Prob > F
Columns	878.1	2	439.052	51.66	0
Error	1504.28	177	8.499		
Total	2382.38	179			

ANOVA TABLE

Source	SS	df	MS	F	Prob > F
Columns	72.6	1	72.6	2.2	0.1431
Error	1911	58	32.95		
Total	1983.6	59			

Fig. 4.15 ANOVA analysis to compare: (**a**) Time Spent by subjects between real, virtual with haptics and virtual without haptics. (**b**) Number of correct lifts when performing the test with the real egg and with the virtual environment with haptics

Time Spent In this case, the time spent lifting the object in each type of test is compared. As shown in Fig. 4.15, the mean time spent lifting a real egg is 6.8 s. In contrast, time needed to lift a virtual egg when the user has force feedback is 10.29 s. If no haptic information is available, the time is 12.13 s.

Figure 4.15 also shows that, in the case of real objects, the data between the 25th and 75th percentiles is represented by a very narrow area. This area is wider in both virtual experiments. However, in the case of virtual experiments without haptics, some outliers are too far from the mean.

An ANOVA analysis was also undertaken to evaluate whether the different kinds of experiments are in fact the same. The result is that the tasks are absolutely different, but in the case of virtual experiments with haptics, the difference with reality is slighter.

Discussions Haptic devices provide human beings with the opportunity to interact with virtual objects or even with remote environments if a slave robot is used. This feeling of immersion within the scenarios can only be achieved if the scenarios effectively simulate how people feel in real life.

This experiment is very suitable for evaluating haptic devices because it allows users to interact with the objects by grasping them. Specifically, the experiment focuses on multi-finger haptic devices that permit interaction with the environment by means of the use of several fingers at once. This experiment compares three different ways of grasping an object: real manipulation, virtual manipulation with haptics and virtual manipulation without haptics

The experiment results show that people can perform this specific task with similar effectiveness whether they are manipulating the real object or the virtual object (with haptics). However, people need more time to complete the task when interacting with the virtual environment. The effectiveness analysis shows that the lack of haptic information makes it really difficult to manipulate virtual objects with accuracy.

4.4.3 Virtual Collaborative Tasks

A multifinger haptic interface like MasterFinger-2 provides two contact points for grasping and manipulating virtual objects; therefore, it is a suitable haptic interface for performing collaborative tasks. A simple virtual application was developed to validate the MasterFinger haptic interface during the execution of collaborative tasks. The application allows a virtual prism to be lifted jointly by two persons as shown in Fig. 4.16.

Each person is located in a different room and no form of direct communication can be established between them. They can only coordinate their movements based on the haptic information received during the interactions and the visual information provided by the graphical interface.

4.4.3.1 Methods for This Experiment

In this experiment the two participants collaborate to lift a deformable object. The experiment consists of two tests: One test requires the participants to lift the virtual prism with visual and haptic information. The other test requires the participants to

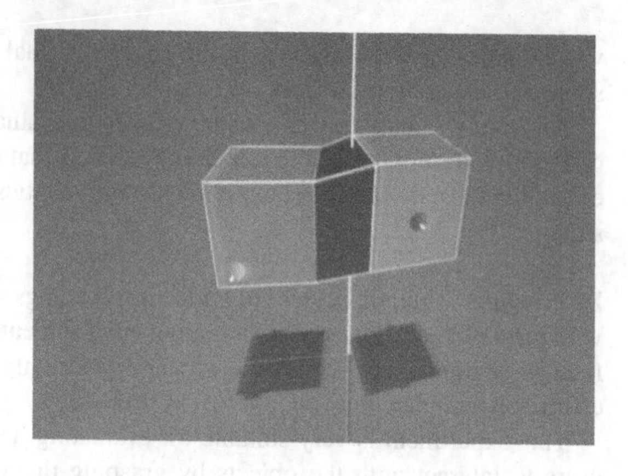

Fig. 4.16 Visual information displayed to both participants in their respective locations

lift the virtual prism with only visual information. If the participants apply too much pressure or pull too force in different directions (more than 3.2 Newtons), the object breaks and the trial is recorded as a failure.

Once the participants are ready to begin the experiment, they are asked to lift the object to a height of at least 20 cm. fifty times and then put it back on the virtual ground.

All of the couples carried out the experiment with and without haptic information. However, each couple carried out each part of the experiment in a different order.

During the experiment, the person who coordinates the experiment registers the most significant parameters that characterize every lift. After every lift, he or she registers whether the experiment was executed correctly or was affected by some incident (for example, the object slipped or broke).

4.4.3.2 Experiment Procedure

Twenty-four UPM students volunteered to participate in the experiment. They were organized into twelve couples to carry out the experiment. Half of the couples did first the collaborative test with haptic and visual information and then the test with only visual information; the other half of the couples did the test the other way around.

The participants were placed in isolated rooms in order to avoid any kind of communication between them; they interacted only through the collaborative scenario. Therefore the participants only received visual information or, depending on which test they were carrying out, haptic information as well.

Each member of the couple used a haptic scenario consisting of a MasterFinger-2 and a graphic screen where they watched a visual representation of the collaborative task. The MasterFinger-2 was mounted upside-down on a metal structure. This position made the MasterFinger workspace larger and the hand position more comfortable. Figure 4.17 shows both participants performing a collaborative lifting; the image also shows the MasterFinger and the visual interface set-up.

Fig. 4.17 Participants performing a collaborative lifting of a prism. Each person is placed in an isolated room

4.4.3.3 Experiment Results

The study undertaken compares the most significant parameters describing the quality of both types of collaborative tasks (haptic and non-haptic). First of all, it considers the learning time necessary for a participant to become accustomed to the task at hand. Secondly, an effectiveness analysis compares the success rate of a task executed using haptic information or only visual information. Finally, the time spent to complete the task was also analyzed.

Success Rate of the Collaborative Task The collaborative task manipulates a virtual prism that can be broken if improperly manipulated, as was described in previous sections. Therefore, the effectiveness of a collaborative task is measured according to the number of successful completions, broken objects and slipped objects.

Figure 4.18 compares both types of collaborative tasks (a collaborative task with haptic data versus a collaborative task with only visual data). The results depicted in this figure show that there is a larger percentage of success (92.5 %, if the collaborative task is carried out with haptic data, as compared to a task carried out with only visual information (83.3 %). Additionally, a visual collaborative task has more than double the percentage of broken objects (15.8 %) as compared to a haptic collaborative task (6.3 %). The percentage of slipped objects is almost negligible for both kinds of collaborative tasks. This proves that haptic data is a key stimulus in enabling a collaborative task to be performed more precisely and efficiently.

An ANOVA test was also performed to compare the two kinds of collaborative tasks. This test compares the number of successful trials. According to these results, a cooperative task with haptics is much more effective and precise than a visual one. The mean number of successful trials for a haptic task is 46.25, while it is 41.67 for a visual task; the mean difference is 4.58, which is significant. Besides, as shown in the plot, the whisker range values of the visual task are very broad. This means that there is a significant dispersion in results, or, in other words, that some participants had serious difficulties performing the task properly. They did not feel comfortable with the task and they did not get used to the visual information.

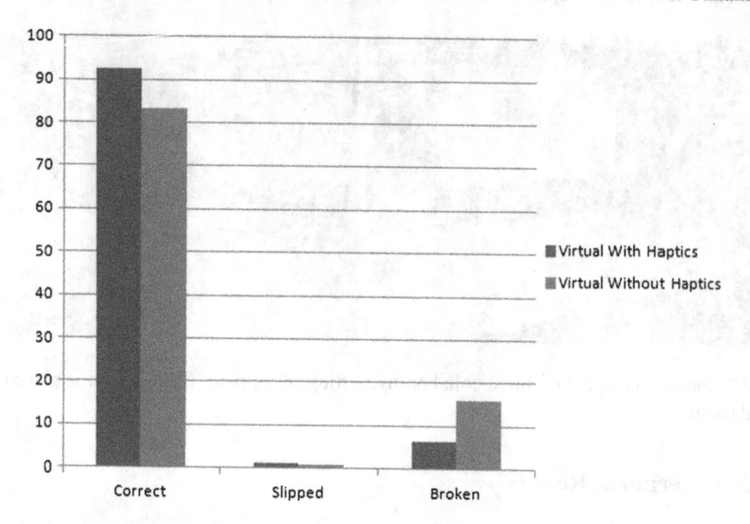

Fig. 4.18 Percentage of successful completions, broken objects and slipped objects in a haptic and visual collaborative manipulation task

Time Spent Participants who carried out the experiment were not used to haptic or human computer interfaces and thus needed some time to become accustomed to the interface and the task.

An analysis of the time taken by subjects to perform this task demonstrates that, after fifteen attempts, the time needed to perform the task stabilizes. This is the learning time. The mean time for performing this task is 9.98 s. Similarly, the evolution of the time needed to perform a collaborative task, but this time with only visual information. An examination of this plot shows that the learning time is twenty attempts. After this time, the mean of each trial converges at 11.66 s. Although the mean time converges, standard deviation shows how significant values are maintained for the duration of the test and can also increase for some trials. This means that there is a significant variability in the time needed to perform the task throughout the experiment. This suggests that the participants do not feel completely comfortable and confident performing this task with only the visual information.

Figure 4.19 shows the results obtained. An analysis of the results shows that the mean time spent on a haptic collaborative task is 9.989 s, whereas the mean time spent on a visual collaborative task is 11.664 s. There is a significant mean time difference between the two types of tasks, 1.675 s. This indicates that haptic stimuli are a key factor in performing the experimental task with more precision and agility. Haptic stimuli make participants' sensations more like those perceived in a real collaborative task.

Discussion This experiment compares haptic and visual collaborative tasks in order to assess the importance of haptic stimuli in this type of task. The experiment conducted and its results clearly demonstrate that haptic stimuli significantly enhance execution of a virtual collaborative task.

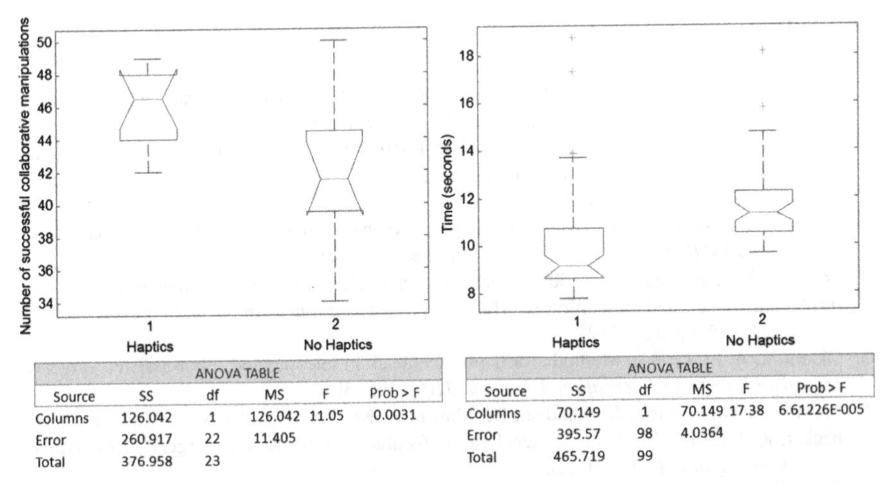

ANOVA TABLE

Source	SS	df	MS	F	Prob > F
Columns	126.042	1	126.042	11.05	0.0031
Error	260.917	22	11.405		
Total	376.958	23			

ANOVA TABLE

Source	SS	df	MS	F	Prob > F
Columns	70.149	1	70.149	17.38	6.61226E-005
Error	395.57	98	4.0364		
Total	465.719	99			

Fig. 4.19 (**a**) Test that compares the number of successful collaborative tasks performed with haptic versus with only visual information. (**b**) Test comparing the mean time of both collaborative tasks performed (haptic versus visual)

The tests performed also demonstrate that people need haptic interfaces that provide stimuli as similar as possible to the stimuli perceived in real interactions. This is confirmed by the fact that the haptic test conducted shows better performance than the test conducted with only visual information.

An analysis of the learning time demonstrates that haptic data tests had a shorter learning time. This serves to confirm that the more similar the stimuli are to a real collaborative interaction, the shorter the learning time required for performing a task.

4.5 Conclusion

In this chapter, a multi-finger haptic device that allows the user to interact with his index, thumb and middle fingers is described. The mechanical design is based on a modular structure in which each module is dedicated to a finger, with redundant axes in order to achieve a high workspace, low inertia, high forces, etc. Software and hardware requirements for real-time control architecture for haptic systems that involve multi-finger haptic interaction are analyzed and the solution designed for this system is described.

Experiments show that the developed 2-finger, and 3-finger haptic devices are suitable for advanced manipulations such as medical simulators, collaborative manipulation of virtual objects or dexterous manipulation of fragile objects.

Acknowledgements This work has been partially supported by Madrid Community in the framework of The IV PRICIT through the project TECHNOFUSION(P2009/ENE/1679), the TEMAR project under grant DPI2009-12283 from the Spanish Ministry of Science and Innovation (MICINN) and UPM under 'Formaci6n de Personal Investigador'.

References

1. Salisbury, J. K., & Srinivasan, M. A. (1997). Phantom based haptic interaction with virtual objects. *IEEE Computer Graphics and Applications, 17*(5), 6–10.
2. Škorc, G., Zapušek, S., Čas, J., & Šafarič R. (2010). Virtual user interface for the remote control of a nanorobotic cell using a haptic-device. *Strojniški Vestnik—Journal of Mechanical Engineering, 56*(7–8), 423–435.
3. Peer, A., & Buss, M. (2008). A new admittance-type haptic interface for bimanual manipulations. *IEEE/ASME Transactions on Mechatronics, 13*(4), 416–428.
4. Waldron, K. J., & Tollon, K. (2003). Mechanical characterization of the immersion corp. haptic, bimanual, surgical simulation interface. In *8th international symposium on experimental robotics* (Vol. 5, pp. 106–112).
5. Okamura, A. M. (2004). Methods for haptic feedback in teleoperated robot-assisted surgery. *Industrial Robot: An International Journal, 31*(6), 499–508.
6. McMahan, W., Gewirtz, J., Standish, D., Martin, P., Kunkel, J. A., Lilavois, M., & Kuchenbecker, K. J. (2011). Tool contact acceleration feedback for telerobotic surgery. *IEEE Transactions on Haptics, 4*(3), 210–220.
7. Ferre, M., Galiana, M., Wirz, R., & Tuttle, N. (2011). Haptic device for capturing and simulating hand manipulation rehabilitation. *IEEE/ASME Transactions on Mechatronics, 16*(5), 808–815.
8. Strolz, M., Groten, R., Peer, A., & Buss, M. (2011). Development and evaluation of a device for the haptic rendering of rotatory car doors. *IEEE Transactions on Industrial Electronics, 58*(8), 3133–3140.
9. Coles, T. R., Meglan, D., & John, N. W. (2011). The role of haptics in medical training simulators: a survey of the state of the art. *IEEE Transactions on Haptics, 4*(1), 51–66.
10. Robotiq Adaptive Gripper—ROBOTIQ flexible robot grippers designer and manufacturer. http://robotiq.com/en/products/adaptive-robot-gripper.
11. Raju, G. J., Verghese, G. C., & Sheridan, T. B. (1989). Design issues in 2-port network models of bilateral remote teleoperation. In *IEEE international conference on robotics and automation* (pp. 1317–1321).
12. Yokokohji, Y., & Yoshikawa, T. (1992). Bilateral control of master-slave manipulators for ideal kinesthetic coupling-formulation and experiment. *IEEE Transactions on Robotics and Automation, 10*(5), 605–620. doi:10.1109/70.326566.
13. Lawrence, D. A. (1993). Stability and transparency in bilateral teleoperation. *IEEE Transactions on Robotics and Automation, 9*(5), 624–637. doi:10.1109/70.258054.
14. Lee, S. S., & Lee, J. M. (2003). Design of a general purpose 6-DOF haptic interface. *Mechatronics, 13*, 697–722.
15. O'Malley, M., & Goldfarb, M. (2002). The effect of force saturation on the haptic perception of detail. *IEEE/ASME Transactions on Mechatronics, 7*(3), 280–288.
16. López, . J., Breñosa, J., Galiana, I., Ferre, M., Giménez, A., & Barrio, J. (2012). Mechanical design optimization for multi-finger haptic devices applied to virtual grasping manipulation. *Strojniški Vestnik—Journal of Mechanical Engineering, 58*(7–8), 431–443.
17. Giachritsis, C. D., Ferre, M., Barrio, J., & Wing, A. (2011). Unimanual and bimanual weight perception of virtual objects with a new multi-finger haptic interface. *Brain Research Bulletin, 85*(5), 271–276.
18. Cervantes-Sánchez, J. J., Hernández-Rodríguez, J. C., & Rendón-Sánchez, J. G. (2000). On the workspace, assembly configurations and singularity curves of the RRRRR-type planar manipulator. *Mechanism and Machine Theory, 35*, 1117–1139.
19. Garcia-Robledo, P., Ortego, J., Ferre, M., Barrio, J., & Sanchez-Uran, M. A. (2011). Segmentation of bimanual virtual object manipulation tasks using multifinger haptic interfaces. *IEEE Transactions on Instrumentation and Measurement, 60*(1), 69–80.
20. García-Robledo, P., Ortego, J., Barrio, J., Galiana, I., Ferre, M., & Aracil, R. (2009). Multifinger haptic interface for bimanual manipulation of virtual objects interaction between two

hands and virtual objects with MasterFinger. In *IEEE international workshop on haptic audio visual environments and games*, Lecco (pp. 30–35).
21. Liu, X. J., Wang, J., & Pritschow, G. (2006). Performance atlases and optimum design of planar 5R symmetrical parallel mechanisms. *Mechanism and Machine Theory*, *41*, 119–144.
22. Cobos, S., Ferre, M., Sánchez-Urán, M. A., Ortego, J., & Aracil, R. (2010). Human hand descriptions and gesture recognition for object manipulation. *Computer Methods in Biomechanics and Biomedical Engineering*, *13*(3), 305–317.
23. Cutkosky, M. R. (1989). On grasp choice, grasp models, and the design of hands for manufacturing tasks. *IEEE Transactions on Robotics and Automation*, *5*(3), 269–279.
24. Endo, T., Kawasaki, H., Mouri, T., Ishigure, Y., Shimomura, H., Matsumura, M., & Koketsu, K. (2011). Five-fingered haptic interface robot: HIRO III. *IEEE Transaction on Haptics*, *4*(1).
25. Hannaford, B., & Okamura, M. (2008). Haptics. In B. Siciliano & O. Khatib (Eds.), *Handbook of robotics* (p. 720). Berlin: Springer. ISBN: 9789-3-540-23957-4. Chap. 20.
26. Cerrada, P., Breñosa, J., Galiana, I., López, J., Ferre, M., Giménez, A., & Aracil, R. (2011). Optimal mechanical design of modular haptic devices. In *Advanced intelligent mechatronics, IEEE/ASME international conference, AIM2011*, Budapest, Hungary.
27. Garcia-Robledo, P., Ortego, J., Barrio, J., Galiana, I., Ferre, M., & Aracil, R. (2009). Multifinger haptic interface for bimanual manipulation of virtual objects. In *IEEE international workshop on haptic audio visual environments and games*. HAVE.
28. Ferre, M., Galiana, I., & Aracil, R. (2011). Design of a lightweight, cost effective thimble-like sensor for haptic applications based on contact force sensors. *Sensors*, *11*, 11495–11509.
29. Galiana, I., Bielza, M., & Ferre, M. (2010). Estimation of normal and tangential manipulation forces by using contact force sensors. In *Lecture notes in computer science springer, eurohaptics 2010*, Amsterdam, The Netherlands.
30. Monroy, M., Ferre, M., Barrio, J., Eslava, V., & Galiana, I. (2009). Sensorized thimble for haptic applications. In *IEEE international conference on mechatronics 2009*, Málaga, Spain.
31. Wirz R, ., Marin, R., Ferre, M., Barrio, J., Claver, J. M., & Ortego, J. (2009). Bidirectional transport protocol for teleoperated robots. *IEEE Transactions on Industrial Electronics*, *56*(9), 3772–3781.
32. National Instruments. Available online on October 2012: http://www.ni.com/pxi/.
33. Melder, N., Harwin, W., & Sharkey, P. (2003). Translation and rotation of multi-point contacted virtual objects. In *Proceedings of the WorldHaptics conference* (pp. 218–277).
34. Nourian, S., Shen, X., & Georganas, N. D. (2006). XPHEVE: an extensible physics engine for virtual environments. In *Canadian conference on electrical and computer engineering, CCECE'06*, May 2006 (pp. 1546–1549).
35. Tuttle, N., & Jacuinde, G. (2011). Design and construction of a novel low-cost device to provide feedback on manually applied forces. *The Journal of Orthopaedic and Sports Physical Therapy*, *41*, 174–179.
36. Lee, M., Gal, J., & Herzog, W. (2000). Biomechanics of manual therapy. In Z. Dvir (Ed.), *Clinical biomechanics* (pp. 209–237). Philadelphia: Churchill Livingstone.
37. Tholey, G., & Desai, J. P. (2007). A general-purpose 7 DOF haptic device: applications toward robot-assisted surgery. *IEEE/ASME Transactions on Mechatronics*, *12*(6), 662–669.
38. Werner, D., Kozin, S. H., Brozovich, M., Porter, S. T., Junkin, D., & Seigler, S. (2003). The biomechanical properties of the finger metacarpophalangeal joints to varus and valgus stress. *The Journal of Hand Surgery (American Volume)*, *28*, 1044–1051.
39. Samur, E., Wang, F., Spaelter, U., & Bleuler, H. (2007). Generic and systematic evaluation of haptic interfaces based on testbeds. In *IEEE/RSJ international conference on intelligent robots and systems, IROS 2007*, October 2007 (pp. 2113–2119).
40. Fisher, R. (1970). Intraclass correlations and the analysis of variance. In *Statistical methods for research workers* (14th ed.). Edinburgh: Oliver and Boyd.

Chapter 5
HIRO: Multi-fingered Haptic Interface Robot and Its Medical Application Systems

Haruhisa Kawasaki, Takahiro Endo, Tetuya Mouri, Yasuhiko Ishigure, and Vytautas Daniulaitis

Abstract This chapter presents the design and characteristics of a five-fingered haptic interface robot named HIRO and its medical application systems. The aim of the development of HIRO is to provide a high-precision three-directional force at the five human fingertips. HIRO consists of a 15-degrees-of-freedom (DOF) haptic hand, a 6-DOF interface arm, and a control system. The haptic interface can be used in a large workspace and can provide multipoint contact between the user and a virtual environment. Three medical application systems using HIRO, a hand rehabilitation support system, a medical training system using plural devices, and a breast palpation training system, are introduced. Furthermore, a hand haptic interface for an advanced palpation training system, which consists of a multi-fingered haptic interface for fingertips and 1-dimensional force display for finger pads, is presented. These systems show the great potential of HIRO.

5.1 Introduction

Most human interfaces in virtual environments use information from visual and auditory sensations. With the addition of tactile sensations to these human interfaces, a human operator can handle objects with a realistic tactile sense and can perform complex tasks in a virtual reality environment. Multi-fingered haptic interfaces that allow multipoint contact between users and a virtual environment have greater potential for various applications than do single-point haptic interfaces. The multipoint interaction allows a user to perform natural actions such as grasping, manipulation, and exploration of virtual objects, and such interaction will dramatically increase the believability of the haptic experience [1, 2].

H. Kawasaki (✉) · T. Endo · T. Mouri
Gifu University, 1-1 Yanagido, Gifu 501-1193, Japan

Y. Ishigure
Marutomi Seiko Co., Ltd., Kurachi Aza-Ikuda 3147-7, Seki-shi, Gifu 501-3936, Japan

V. Daniulaitis
UAB Aksonas, Europos pr. 121-305, 46339 Kaunas, Lithuania

I. Galiana, M. Ferre (eds.), *Multi-finger Haptic Interaction*,
Springer Series on Touch and Haptic Systems, DOI 10.1007/978-1-4471-5204-0_5,

Haptic interfaces that present force and tactile feeling to the fingertips of humans have been investigated extensively over the last two decades [3–13]. Some of these interfaces [10–13] provide three-dimensional (3D) force feelings to the human fingertip, and a few haptic interfaces cover the workspace of the human arm. These interfaces are classified into two types: wearable-type interfaces [6], which are mounted firmly on the human hand, and grounded-type interfaces [11–13], which are located in positions that match and are opposed to the human hand. The wearable-type haptic interface has a large workspace, but with this interface it is difficult to present three-directional forces to the user or simulate the weight of virtual objects on the fingertips, because the hand mechanism is mounted on the back of a human hand, and the exerted force is only a one-directional force. In contrast, the grounded-type haptic interface generally has a fairly small workspace compared with the wearable-type haptic interface, but a grounded-type haptic interface consisting of an arm and fingertips could be used in a large workspace. Here note that an encountered-type haptic interface [14, 15] was proposed in addition to the above mentioned classification. This interface would be an ideal haptic interface, but it has proven to be difficult to achieve using the present technology. A haptic interface must be safe, function in a wide operation space, and represent the weight of virtual objects as well as force at the contact points. In addition, it should not have an oppressive feeling when it is attached to a person, and it should not represent its own weight.

Medical application systems such as rehabilitation support [16, 17], surgery training [18, 19], and palpation training [20, 21] have attracted attention as effective uses of the multi-fingered haptic interface. In most real situations, humans manipulate objects using force and move in response to tactile feelings on their fingertips, finger pads, and palm. For example, medical doctors search for tumors during a breast palpation and manipulate internal organs during surgery using not only their fingertips but also their finger pads and palm. To allow doctors to practice such medical procedures in a virtual environment, a hand haptic interface that can apply forces to not only the fingertips but also the finger pads and palm is desired.

This chapter introduces a developed multi-fingered haptic interface robot called HIRO [13, 22] and its medical application systems. HIRO, which consists of a 15-degrees-of-freedom (DOF) haptic hand, a 6-DOF interface arm, and a control system, can present high-precision three-directional force at the five human fingertips. Attempts have been made to apply HIRO to a hand rehabilitation support system [23], a medical surgery training system using plural devices [24–26], and a breast palpation system [27, 28]. An outline of each of these systems is introduced here. Furthermore, a hand haptic interface [29, 30] that consists of 1D force display devices for the finger pads and palm and a 3D haptic interface for the fingertips is introduced. These medical application systems show the high potential of the multi-fingered haptic interface robot HIRO, and the hand haptic interface shows expanding developments in the multi-fingered haptic interface.

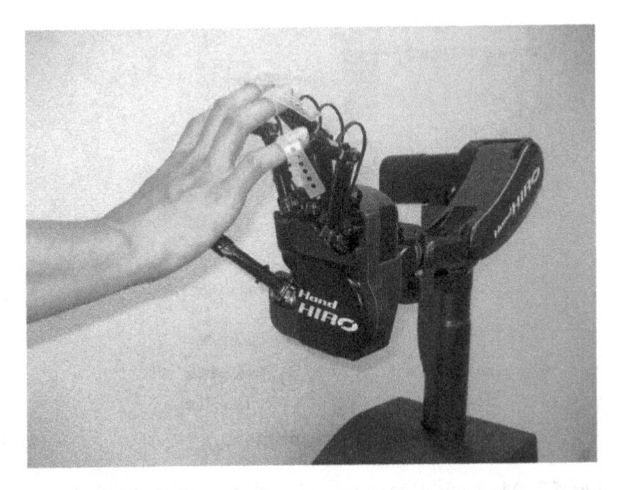

Fig. 5.1 Multi-fingered haptic interface robot, HIRO III

5.2 Hand HIRO

Multi-fingered haptic interfaces joined to the arm can provide a wide operation space. However, most of these interface systems are mounted on the back of the human hand (see, for example, the CyberForce [6]). Fixing the haptic interface to the hand gives the operator an oppressive feeling, since the interface is firmly bound to the human hand. To reduce the oppressive feeling and increase safety, we developed a new Haptic Interface RObot named HIRO [13, 22] to represent force feeling to all fingers of the human hand. HIRO can present a three-directional force at each of the five human fingertips. It is placed opposite a human hand, and the haptic fingertips are connected to the human fingertips through finger holders and passive spherical magnet joints. This section presents the mechanical design of the most recent version of HIRO, called HIRO III [22]. In addition, HIRO III is available commercially under the name Hand HIRO through UAB Aksonas [31] and Marutomi Seiko Co., Ltd. [32].

5.2.1 Mechanical Structure

HIRO III consists of an interface arm, a haptic hand with five haptic fingers, and a controller, as shown in Fig. 5.1. When the operator moves his/her hand, HIRO III follows the motion of the operator's fingers and represents the force feeling. The operator feels just a slightly oppressive feeling, because the part connecting the human hand and the haptic interface is limited to his/her fingertips.

The mechanical structure of the haptic finger is shown in Fig. 5.2(1). It has 3 joints allowing 3 DOF. The first joint, relative to the hand base, allows abduction/adduction. The second and third joints allow flexion/extension. All joints are driven by DC servomotors with gear transmissions and rotary encoders. A workspace of the haptic finger is shown in Fig. 5.2(2). The thumb is designed

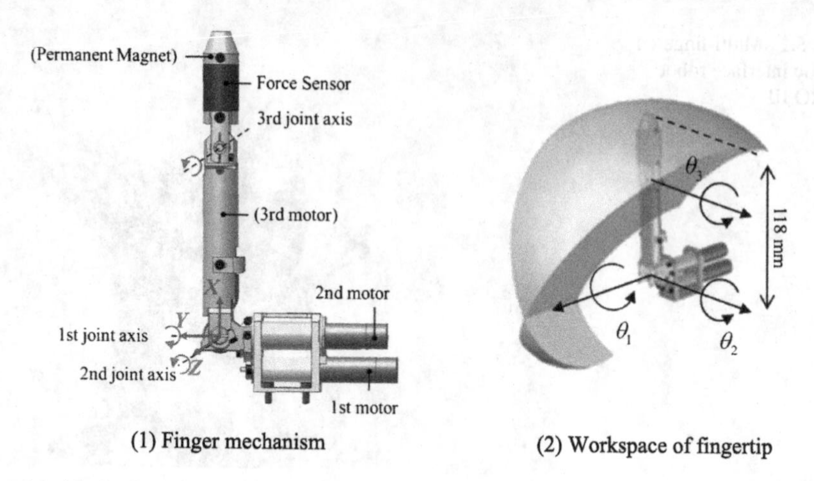

(1) Finger mechanism (2) Workspace of fingertip

Fig. 5.2 Mechanical structure and workspace of the haptic finger

Fig. 5.3 Finger holder

to work in a wide space, because the workspace of the human thumb is larger than that of the other fingers. The finger layout of the haptic hand is designed to maximize the volume formed by intersecting workspaces of the haptic and human fingers. The haptic index finger was designed based on statistical data for Japanese males [13]. The position of fingers in the haptic hand was designed based on a consideration of this geometrical relation.

To read the finger loading, the 3-axes force sensor is installed in the second link of each haptic finger. To manipulate the haptic interface, the user has to wear a finger holder on his/her fingertips, as shown in Fig. 5.3. This finger holder has an iron sphere which forms a passive spherical joint when attached to the permanent magnet at the force sensor tip. The passive spherical joint has two roles. One role is adjusting for the differences between the orientation of the human and haptic fingers. Each human finger has 6 DOF, while each haptic finger has 3 DOF. Hence, an additional passive 3 DOF are needed. The second role of the spherical joint is to ensure that the operator can remove his fingers from the haptic interface when it malfunctions. The suction force by the permanent magnet is 4.3 [N]. The maximum output force at a fingertip is 3.6 [N]. The maximum fingertip velocities v_x, v_y, and v_z are 0.43, 0.43, and 0.23 [m/s], respectively. Table 5.1 shows specifications of the HIRO III hand.

The interface arm with 6 DOF is designed such that it is compact and covers the workspace of human desk work. The lengths of the upper arm and the forearm are

Table 5.1 Specifications of the HIRO III hand

	Component	Quantity
Hand	Number of fingers	5
	Degrees of freedom	15 DOF
	Weight [kg]	0.78
Finger	Degrees of freedom	3 DOF
	Weight [kg]	0.12
	Maximum output force [N]	3.6
	Maximum pulling force between the magnet and the metal ball [N]	4.3
	Volume of workspace [cm^3]	705 (thumb)
		587 (other)

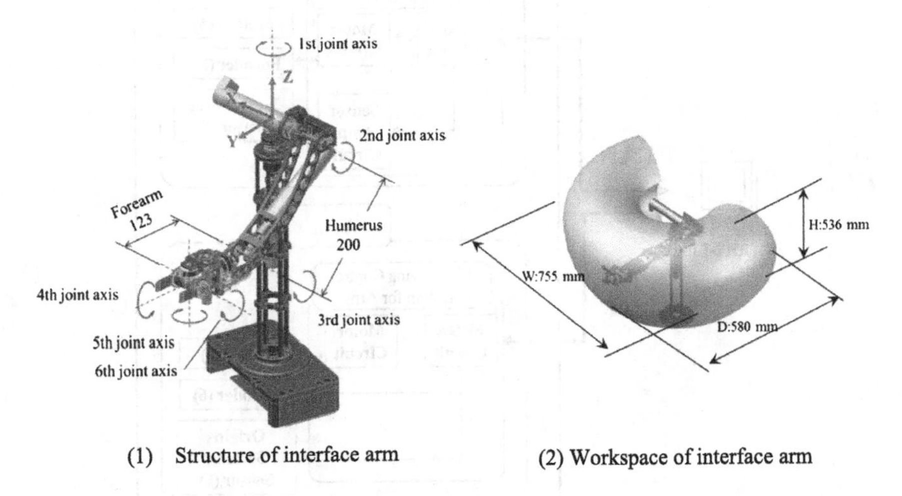

Fig. 5.4 Structure and workspace of interface arm

0.2 and 0.17 [m], respectively. The arm joints are actuated by servomotors equipped with rotary encoders and gear transmissions. As a haptic interface, the combination of the arm and hand makes a wide workspace, as shown in Fig. 5.4. Table 5.2 shows specifications of the HIRO III arm.

5.2.2 Control System

The control system of HIRO III is shown in Fig. 5.5. To present the force at the five human fingertips, 15 DC motors and five force sensors are installed in the haptic hand. As a result, the communication cable between the control PC and the haptic hand consists of 60 wires for the 15 encoders, 30 wires for the 15 motors, and

Table 5.2 Specifications of the HIRO III arm

Degrees of freedom		6 [DOF]
Weight		under 3 [kg]
Maximum output force		over 56 [N]
Workspace		0.09 [m^3]
Operating angle of joints	1st/2nd/3rd	−110∼110/−125∼0/0∼145 [deg]
	4th/5th/6th	−90∼90/−45∼45/−60∼60 [deg]

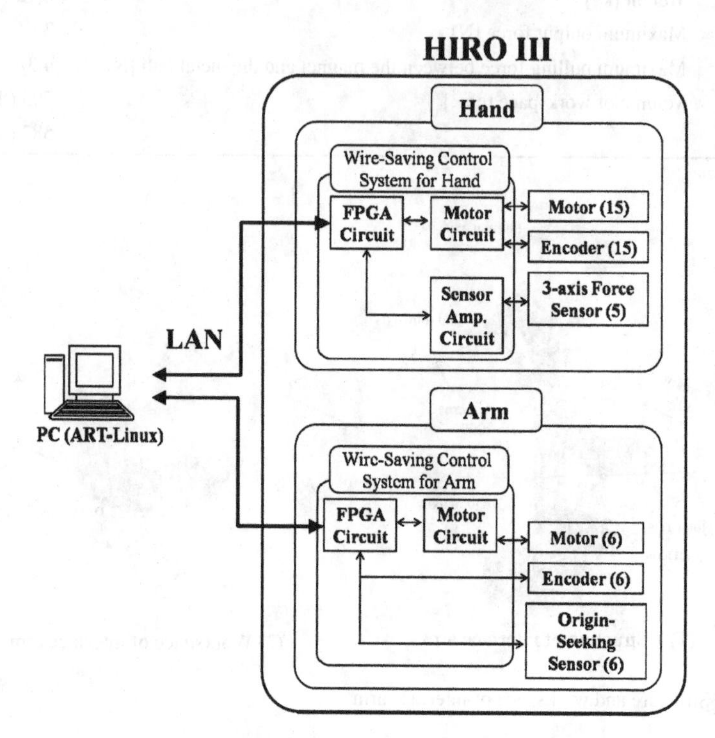

Fig. 5.5 Control system of HIRO III

40 wires for the five force sensors. Similarly, the communication cable between the control PC and the haptic arm consists of 102 wires for the six encoders, motors, and origin-seeking sensors. Therefore, the total number of wires between the haptic interface and the control PC is 202. These wires greatly obstruct the smooth movement of the haptic interface. We have developed a wire-saving control system for the haptic hand which consists of the interface FPGA (Field-Programmable Gate Array) circuit, the force sensor amplifier circuit, and the motor amplifier circuit. These circuits are mounted in the haptic hand frame, as shown in Fig. 5.6. In the FPGA, the VHSIC (Very High Speed Integrated Circuits) Hardware Description Language (VHDL) was used to create a counter driver to count the up/down pulses from the motor amplifier circuit, a Pulse Width Modulation (PWM) driver to pro-

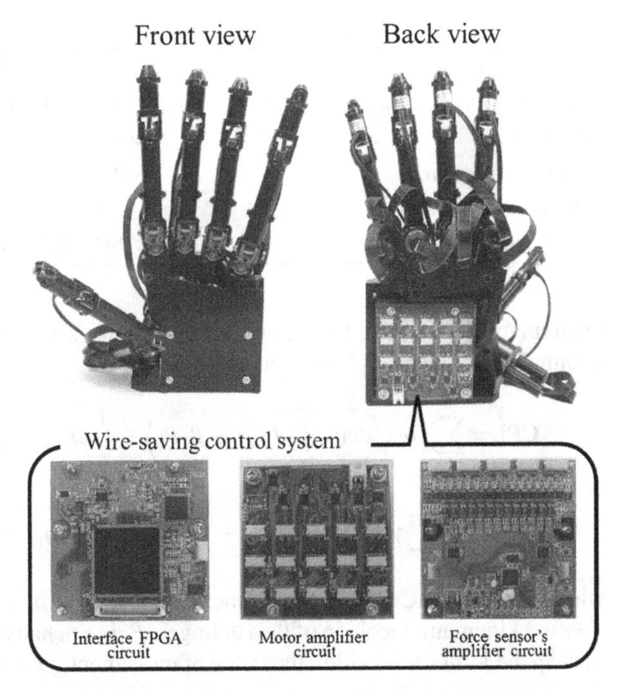

Fig. 5.6 Haptic hand in which the wire-saving control system is installed inside the handbase

vide the PWM outputs of 15 motors, and a sensor driver to read the values of the five force sensors and to provide the zero adjustment value of the five force sensors. The FPGA sends a PWM signal to the motor amplifier circuit when it receives the duty ratio signal from the control PC.

The 3-axes force sensor [33] is installed in the second link of each haptic finger and is connected to the sensor amplifier circuit. By using the strain gauge in the pressure port of the sensor element, the force sensor measures the force from the deformation of the metal beam. The reasons for using the strain gauge are: (1) the sensor has good linearity, (2) the effect of the temperature change is small, and (3) the force sensor is suitable for the measurement of dynamic phenomena because the amount of displacement is extremely small. The specifications of the 3-axis force sensor are shown in Table 5.3.

5.2.3 *Interface Control*

The haptic hand posture is non-unique when the five fingertip positions are fixed, because of the redundancy of HIRO. To take the redundancy into account, the hand manipulability-based control is adopted. This control method combines finger force control and arm position control. Each finger is controlled independently by a PI (Proportional and Integral) force control using a force error at the fingertips. In arm position control, a desired hand posture is determined to maximize the hand

Table 5.3 Specifications of the Force Sensor

Rated capacity	5	[N]
Rated output (X-, Y-, Z-direction)	±0.3, ±0.3, ±0.1	[mV/V]
Safe overload	400	[%]
Nonlinearity, hysteresis	$\leq \|1.0\|$	[%RO]
Repeatability	≤ 1.0	[%RO]
Size/Weight	$\Phi 14 \times H27$ [mm]/12.8 [g]	

manipulability measure to respond to an operator's various hand poses. The hand manipulability measure is defined as follows:

$$\mathrm{CPI} = \sum_{i=1}^{5}\left(\alpha_i\sqrt{\det\left(\boldsymbol{J}_{Fi}^T\boldsymbol{J}_{Fi}\right)} + \beta_i P_i\right) - \frac{1}{2}(\boldsymbol{q}_{Ad} - \boldsymbol{q}_A)^T\boldsymbol{\Gamma}(\boldsymbol{q}_{Ad} - \boldsymbol{q}_A) \quad (5.1)$$

$$P_i = -\sum_{i}^{3}\gamma_i\left[\exp\left\{-\mu(q_{ij} - a_{ij})\right\} + \exp\left\{\mu(q_{ij} - b_{ij})\right\}\right] \quad (5.2)$$

where CPI means Control Performance Index, α_i and β_i are weighting coefficients, $\boldsymbol{J}_{Fi}$ is a kinematic Jacobian of the ith finger, P_i is a penalty function of the ith finger to keep the joint angle within the range of movement, γ_i is the weighting coefficient, μ is the parameter to adjust an exponential function, a_{ij} and b_{ij} are the lower and upper limits of the jth joint angle of the ith finger, respectively, $\boldsymbol{q}_A \in R^6$ is the current arm joint angle vector, $\boldsymbol{q}_{Ad} \in R^6$ is a desired arm joint angle vector, and $\boldsymbol{\Gamma} > 0$ is a weighting matrix. The desired arm joint angle is defined so as to maximize equation (5.1) under a constraint condition in which the five haptic fingertip positions are fixed to the operator's fingertip positions. The second term on the right-hand side in Eq. (5.1) is added to prevent a large arm motion, which sometimes confuses the operator because it creates the illusion that the device is malfunctioning. The optimum arm joint angle is computed by the conjugate gradient method in line with the convergence criterion $\mathrm{CPI}[k] - \mathrm{CPI}[k-1] < \delta$, where k is the iteration number and δ is the scalar constant.

The finger control is the PI force controlled in the movable joint angle and the PD (Proportional and derivative) position control at the limit of the movable joint angle. After reaching the limit of the movable range, the finger that has been switched to the position control is switched back again to the force control when the direction of the joint torque input is the same direction as that of the joint angle apart from the limit of the movable angle. In this force control, the feedback gain is constant. When a user manipulates the haptic interface in free space, the user feels the inertia and friction of the interface, and therefore, a high-gain feedback is desired. On the other hand, when a user uses the haptic interface in the constraint space, namely, when the user contacts a virtual object, the high-gain feedback creates a vibrating response, and a low-gain feedback is desired. Therefore, we improved the force control of the haptic fingers by using a non-constant gain. The controller uses a high gain when the desired force is zero and a low gain when the desired force is not zero. The control PC uses a real-time OS (ART-Linux) to guarantee a 1 [ms] sampling time.

5.3 Medical Application Systems

Medical application systems [16–21] have shown the effectiveness of haptic interfaces, but most of these systems use single-point haptic interfaces. The use of a multi-fingered haptic interface has contributed to the development of novel medical application systems such as a hand rehabilitation support system [23], a surgery training system using plural devices [24–26], and a breast palpation training system [27, 28]. An outline of these systems is presented here.

5.3.1 Hand Rehabilitation Support System

Finger function is very important for daily life, and finger injury can be a very serious problem. Patients with finger injuries need timely and persistent rehabilitation to recover their lost abilities and to resume their normal daily lives. Long rehabilitation training sessions with therapists, who are in relative shortage, are not always available to patients. One solution to this problem could be a robotic finger rehabilitation system used by individual patients. Recently, many robotic systems for rehabilitation have been studied. Many aspects of robotic arm rehabilitation therapy [34, 35], including clinical tests [36], have been reported.

Hand rehabilitation, however, is particularly difficult, because the hand possesses many degrees of freedom of motion, and the device must be relatively small. Connelly et al. [17] have developed a virtual reality (VR) hand rehabilitation system with a pneumatic glove, and Oblak et al. [37] have developed a universal haptic device for arm and wrist rehabilitation. In the former, it is slightly difficult for a patient with hand paralysis to wear the glove, and the latter can provide exercise at the shoulder, arm, and wrist only, not the hand or fingers. Compact devices with a cable system to be used after a stroke [16], and exoskeleton-type robotic devices for hemiplegia patients [38, 39] have been presented. The former is a compact device but provides only limited exercises because of the cable drive with one DOF. The latter can assist not only with flexion/extension but also with abduction/adduction and thumb opposition motions, but its purpose is to assist only with joint motions, not 3-directional fingertip motions.

Surface electromyograms (sEMG) are used to record the electrical activity of muscles underlying the skin and are useful for rehabilitation robotic devices. Rehabilitation control strategies for upper limb rehabilitation [40] and a trial of the recovery process for wrist rehabilitation [41] have been reported. However, the application of sEMG to hand rehabilitation has not been widely researched.

There are many patients with finger paralysis; for example, carpal tunnel syndrome may lead to finger paralysis. These patients need hand rehabilitation to resume their activities of daily life. A solution to this problem would be a rehabilitation system that could provide 3-directional motions to the fingertips. Usually, it is easier for people to imagine 'moving the fingertips' than 'moving each finger joint.' We have developed a hand rehabilitation system using the HIRO, as shown in

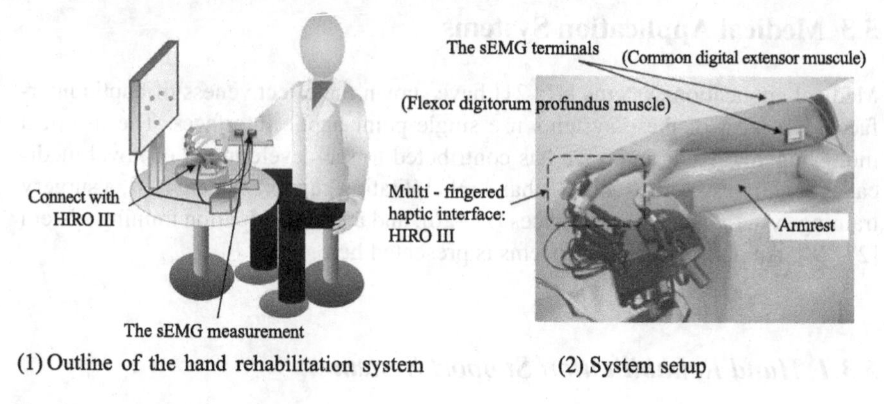

Fig. 5.7 Hand rehabilitation system

Fig. 5.7. To achieve effective rehabilitation for patients, this system utilizes sEMG control. A finger joint angle estimation system that utilizes sEMG has been developed [42], and this system will be helpful for complex exercises. The system will be effective for patients with peripheral nerve injury of the arm. This research could result in a challenging clinical trial, and it is of great significance to integrate the sEMG into a rehabilitation robotics system for the fingers.

5.3.1.1 System Configuration

Our system uses the HIRO hand for the fingertips and sEMG information from the patient's forearm or hand. Since this haptic device is quite small, it can be used on a desk while the patient is seated. First, this system decides the proper fingertip exercises based on the sEMG results; after that, the device provides exercises at the fingertips. For example, if the sEMG from muscles for finger extension (extensor digitorum muscle) activates and crosses over a certain threshold value, the haptic device exerts 3-axis forces at each fingertip to open the fingers.

5.3.1.2 sEMG for Hand Rehabilitation

The sEMG measures bioelectrical signals from muscles in response to voluntary contractions. These signals provide various types of information about a person's intent. It is therefore hoped that the sEMG will function as a new type of human—machine interface, and it has been studied by many researchers [43]. In the present study, the haptic device was controlled by the sEMG. First, the motion intent of the user was extracted from the sEMG, and the fingertip exercises along with the intent were then provided to the user by the HIRO. This biofeedback is very easy to image for the patient compared with waveforms or sounds. And as opposed to completely passive exercise, the proposed system can give the patient a rehabilitative experience that reflects the patient's intent.

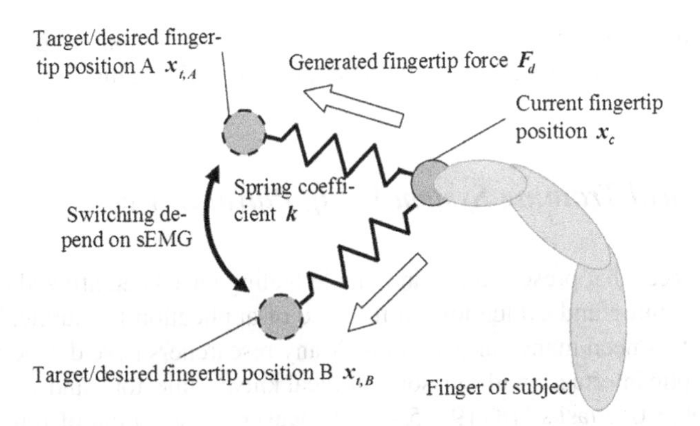

Fig. 5.8 Virtual spring model

5.3.1.3 The Control Strategy Outline

The sEMG is helpful for understanding the patient's intent, but raw sEMG data also provide undesired information (for instance, typical biological noise or personal difference, and on the like), so signal processing will be needed. For signal processing of the sEMG, the system adopts an extracting feature value method, determining the mean absolute value (MAV) or root mean square (RMS) with segmentation and normalizing [44]. For these processes, it will be useful to use feature values rather than raw sEMG data.

After obtaining the desired fingertip exercise by signal processing of the sEMG, the system must transform from the exercise into a continuous fingertip trajectory. The haptic device HIRO provides fingertip forces depending on the desired fingertip trajectory. For rehabilitation, force control is appropriate rather than position control in that the force can be adjusted. As a simple example, one switches the target positions based on the intent. It is necessary to measure these positions beforehand. When the joints of a patient don't have contraction, the target positions can be measured with the help of a therapist or by the patient. When the joints have contraction, the length of the patient's finger link and range of motion are measured, and solving the kinematics with these parameters will identify the target positions. The simplest method to generate the desired fingertip forces is to use the virtual spring model, as shown in Fig. 5.8. For the evaluation results of the developed system, see [42].

As a pretest for the clinical trial, we obtained experimental results with the system for able-bodied subjects. This study targeted two types of movement with two sEMG terminals. The experimental results show that the system provided 3-directional forces at each fingertip corresponding to the subject's sEMG, and that the subjects felt natural movement support. In the future, it will be necessary to modify the system so that it can provide a greater variety of movements. For example, the pinch exercise using the thumb and index finger is very important and is also a barometer for recovery. Paralysis related to this motion can be caused by carpal canal syndrome or injury of the medial nerve. Of particular concern is the fact that Japanese use chopsticks in daily life, and this tool requires a pinch motion. Our intent is to carry out a clinical trial with the proposed system.

5.3.2 Medical Training System Using Plural Devices

Haptic interfaces that present force and tactile feeling have been utilized in the area of medical training and evaluation. In the case of application for surgical training, medical doctors need many surgical tools. Many researchers have developed single tool-type haptic interfaces with scissors, surgical knives, injectors, and so on, for use in limited objective tasks [18, 19, 45–47]. However, presentation of force feelings of multiple surgical tools requires many single tool-type haptic interfaces, and providing many such interfaces would require an appropriate installation location and cost a great deal. Moreover, most single tool-type haptic devices are fixed to a base and thus cannot be displaced. The multi-fingered haptic interface has the advantage of force feeling for plural surgical devices [24].

5.3.2.1 Conceptual Idea

Plural tool-type devices require that each tool-type device be easy to put on and take off, that the overall system be small, and that the system be available for use in a large workspace. These requirements are fulfilled by equipping the multi-fingered haptic interface with tool-type devices. Figure 5.9 shows the concept for presenting force feelings of plural tools. The multi-fingered haptic interface is equipped with exchangeable surgical tool-type devices. The proposed haptic system has the following advantages: (1) Each tool-type device is very simple and in some cases has features that are superior to those of the equivalent conventional devices. (2) It is easy to exchange the plural tool-type devices using permanent magnets at the fingertips. (3) It occupies less space than previous single tool-type haptic systems. (4) The tool-type devices can be used over a wide area. (5) Force and position can be measured and controlled by the multi-fingered haptic interface. With these several advantages, the proposed system has greater potential for the force feelings of plural tool devices than does a single tool-type interface.

5.3.2.2 Cutting Simulation

Many surgical tools are scissors-type devices. Using the scissors tool-type devices with our system, it is easy to express various surgical scissors by changing the knife blade of the scissors in VR. The developed scissors tool-type device is shown in Fig. 5.10. It weighs 70 g. Three iron balls are attached at one supporting point and two working points to connect to the HIRO fingers. As a result of analyzing the workspace of the devices, the system is attached at the thumb, index finger, and ring

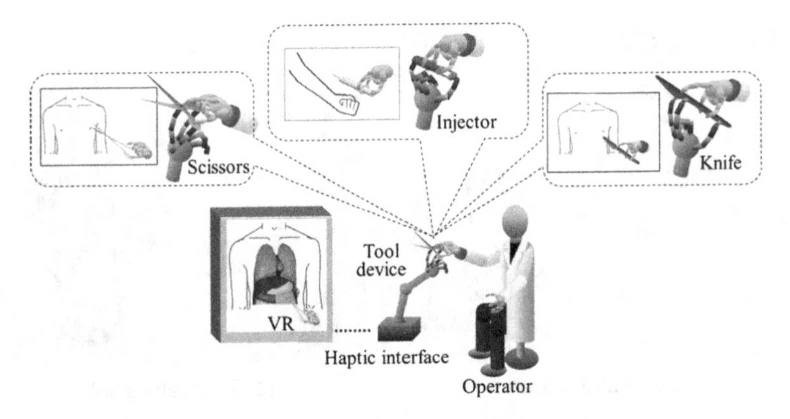

Fig. 5.9 Concept of plural surgical tool devices presentation

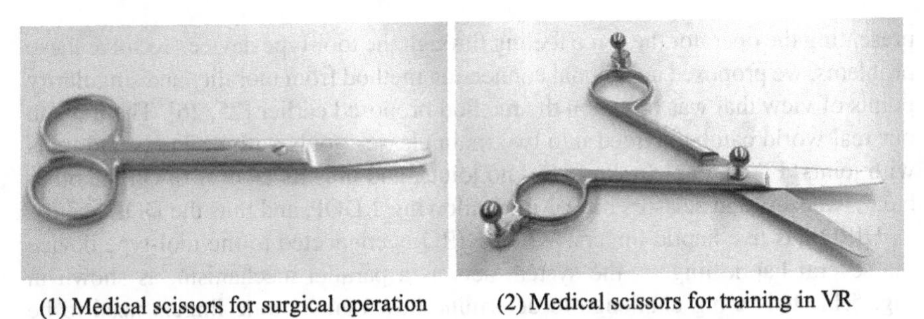

(1) Medical scissors for surgical operation (2) Medical scissors for training in VR

Fig. 5.10 Medical scissors

finger, as shown in Fig. 5.11(1). Figure 5.11(2) shows a task in which an operator cut a virtual sheet using the medical scissors-type device.

A computational method for cutting sheets has been proposed in Ref. [45]. The present system uses this method. Haptic fingers follow the desired forces through PI force feedback control, which are computed in the VR environment. The operator cuts a virtual sheet using virtual scissors with reactive feeling. When the earlier system HIRO II [24] was equipped with a scissors tool-type device, the device displayed the force feeling for the cutting sheet. Experimental results successfully showed the presentation of force feeling depending on the sheet characteristic. However, the error of the force was not small enough for use in surgical training, and thus a new design approach for tool-type devices was proposed [25, 26].

5.3.2.3 Optimum Connection

In our previous research, the connection between the HIRO finger and the tool-type device was determined by trial and error, and its optimal connection was not clarified. Technically clarifying the optimal connection is important with regard to

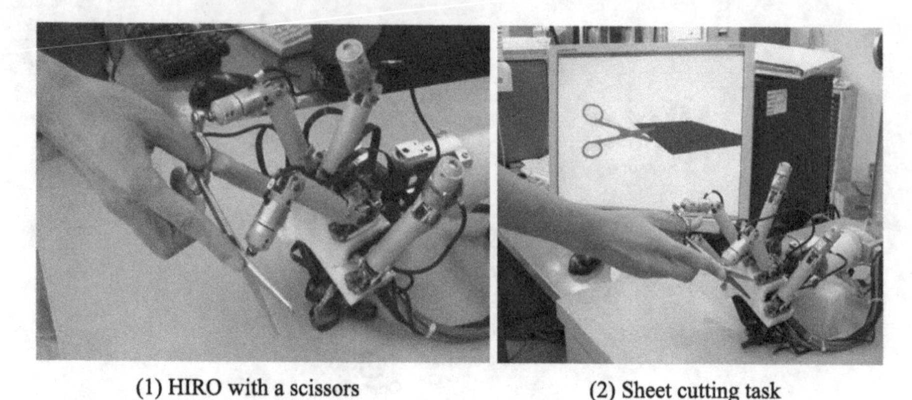

Fig. 5.11 HIRO connected to a medical scissors-type device

presenting the operator the force feeling through the tool-type device. To solve these problems, we proposed an optimal connection method from mobility and singularity points of view that was based on the method proposed earlier [25, 26]. The tools in our real world can be divided into two main classes: tools with no joints and tools with joints. For example, a knife has no joints, and thus its DOF is 6. On the other hand, tweezers and scissors have 1 joint allowing 1 DOF, and thus the DOF is 7.

HIRO has five haptic fingers. When HIRO is connected to the tool-type device via several haptic fingers, the system acts as a parallel mechanism, as shown in Fig. 5.11. It is a big challenge to determine how many haptic fingers need to be connected to a given tool-type device, and which fingers should be used for the connections. These problems are bound up with presenting the operator the force feeling through the tool-type device. To solve these problems, we propose an optimal connection method from mobility and singularity points of view. With regard to mobility, the number of fingers necessary for HIRO to have connected to the tool-type device is determined. Figure 5.12 shows a knife-type device that connects to HIRO optimally. The knife-type device is connected to the thumb, index finger, and little finger of HIRO via a connection link. Figure 5.12(3) shows a task in which an operator manipulated the knife device in the virtual environment. A detailed analysis of this setup and a description of an associated experiment are presented in Refs. [25, 26]. In particular, the optimal connection between HIRO and a knife-type device was studied in [25], and the connection between HIRO and a tweezers-type device was considered in [26]. See [25, 26] for the evaluation results of the developed systems.

5.3.3 Breast Palpation Training System

Despite the potential utility of clinical breast examination (CBE), doctors' palpation skills are often inadequate, and it is difficult to train doctors in this skill. The sense of

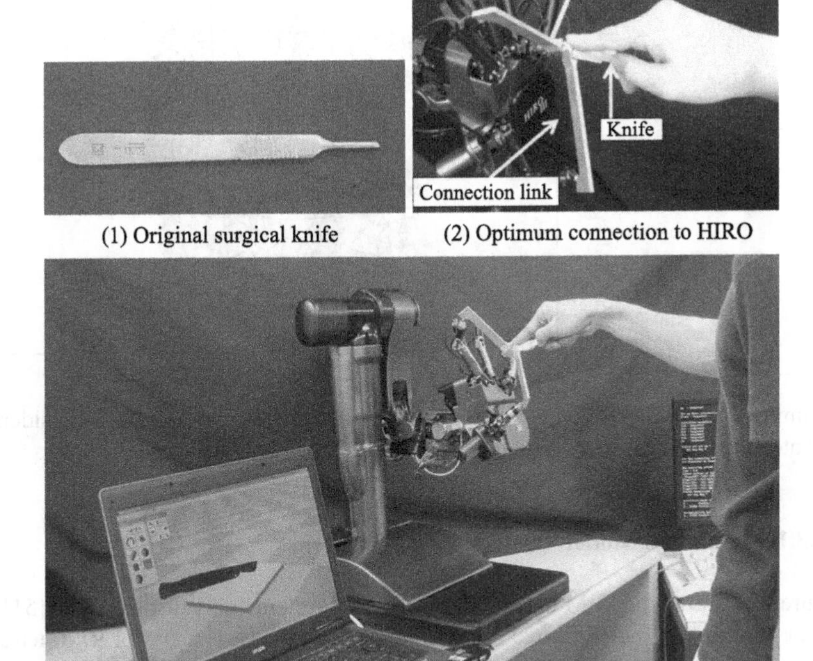

(1) Original surgical knife (2) Optimum connection to HIRO

(3) Manipulation of knife device in virtual environment

Fig. 5.12 Knife-type device

touch is extremely important when training physicians to perform CBE [48]. Breast palpation is a practiced tactile skill that does not necessarily correlate with cognitive knowledge about breast cancer detection. Training that emphasizes tactile skills is known to improve examination sensitivity. With a combination of technologies, such as VR and haptic interfaces, it is possible to greatly extend the capabilities and effectiveness of training simulators [49]. A simulation can record kinematics, touch, and force feedback for later display to a trainee. This technology could be used to train medical students before they palpate a real patient, and could also be used by trained physicians to improve their skill.

This section focuses on the development of a realistic palpation tool for lesions, where the user can touch soft tissue looking for a tumor beneath the surface. In this work a realistic model has been built using the finite element method (FEM), which was fast enough for a relatively complex model in real time. To simulate a medical palpation, the simulation has to handle the deformation caused by multiple fingers contacting the surface at the same time. In multiple fingers contact in Ref. [50], detection and deformation were achieved by computing the reaction force for each finger individually [50]. The fingers during palpation are close to each other, and

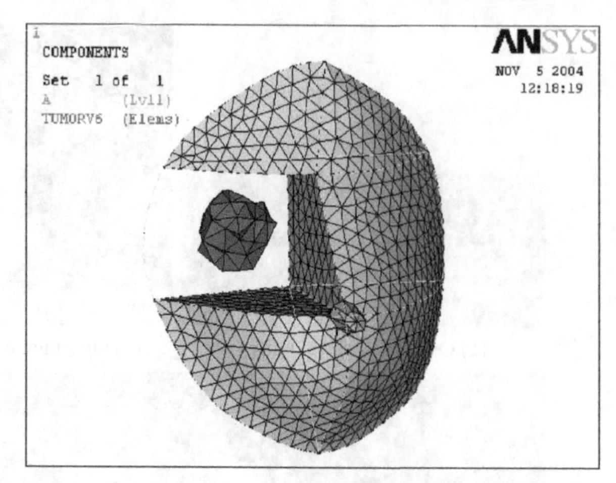

Fig. 5.13 The breast mechanical model with tumor

assuming that no interaction is taking place is not realistic. Our approach considers the interaction that happens between the forces exerted by each finger.

5.3.3.1 Simulation Model

Figure 5.13 shows the FEM in combination with an "elementary displacement [51]" solver applied to a virtual female breast model. The breast consists of 9720 tetrahedra and 2109 nodes. The breast surface is formed of 984 triangular faces and 527 nodes, directly accessible for palpation. This discretization level assures an average FE edge of 6 mm. The elementary displacement approach allows splitting the model processing into two phases, the pre-computation phase (off-line) and the real-time phase (online). In the pre-computation phase, all the possible loading cases for elementary (unitary) displacement are examined, and deformation and equivalent forces at the contact points are stored. According to the displacement induced by the touched nodes of each contact point, the associated forces can be determined in real time. Based on this approach, forces could be computed at 1 KHz. Moreover, the interaction with multiple points of contact (five fingers) has also been realized [28].

5.3.3.2 Palpation Training System

Figure 5.14 shows a developed palpation training system [52]. The user can see a composite image of 3D computer graphics (CG) of the breast and a dynamic picture of his/her hand in the screen. The experiments showed that it is now possible to calculate global deformations and force feedback for the soft object in real time by using a graphic processor array (GPA) system for haptic rendering. However, the deformations and forces processing time when a conventional FE procedure is implemented limits the virtual breast model size to few vertices, so it requires further

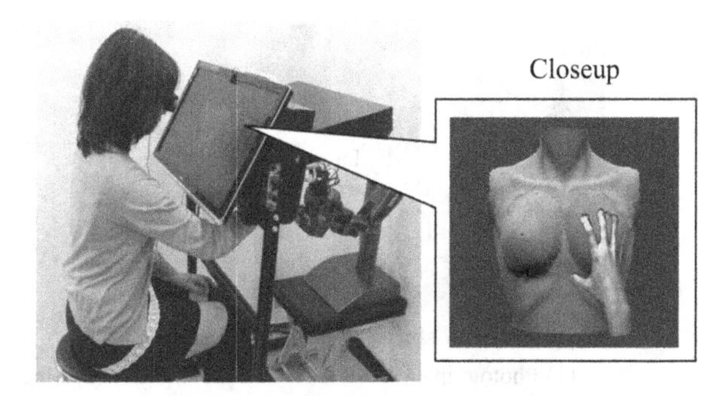

Fig. 5.14 Developed breast palpation training system

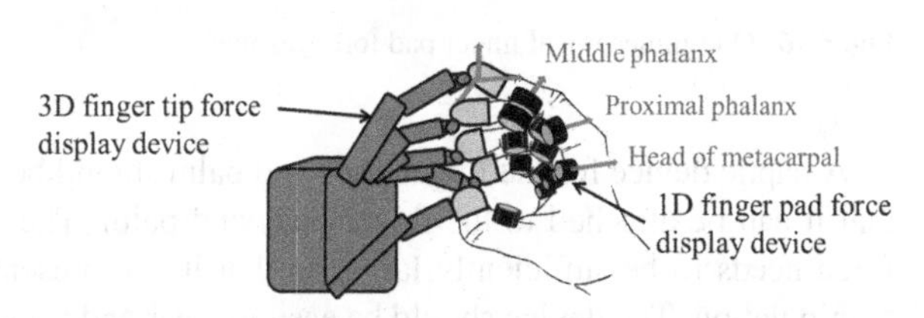

Fig. 5.15 Concept of proposed hand haptic interface

development and improvement. Another issue of the simulation is determination of the parameters of human tissue. Then, a validation method has to be developed to validate the simulation and experiment results.

5.3.3.3 Hand Haptic Interface

In most real situations, humans manipulate objects using force and tactile feelings on their fingertips, finger pads, and palm. For example, medical doctors search for tumors during a breast palpation and manipulate internal organs during surgery using not only their fingertips but also their finger pads and palm. To allow doctors to practice such medical procedures in a virtual environment, a hand haptic interface that can apply forces to not only the fingertips but also the finger pads and palm is desired. Based on the knowledge that the density of tactile sensory organs in the human finger is high in the fingertips and relatively low in the finger pads and palm [53], a hand haptic interface [29, 30] that applies 3D force feelings to the finger tips and 1D force feelings to the finger pads and palm was developed, as shown in Fig. 5.15. In this figure, a human operator is displayed receiving 3D forces at the fingertips through the 3D fingertip force display device and 1D force in the direction normal to the surface of the finger pads and palm through the 1D force display device. The 1D force display devices are attached to the proximal phalange and the metacarpal head of the thumb and the middle phalange, the proximal phalange, and the metacarpal head of the finger. The metacarpal head is on the palm. The total maximum number of displayed points on the finger pads and palm is 14. We have produced a hand haptic interface based on the above-mentioned design concept, in which the multi-fingered haptic interface robot HIRO is used to apply 3D fingertip forces.

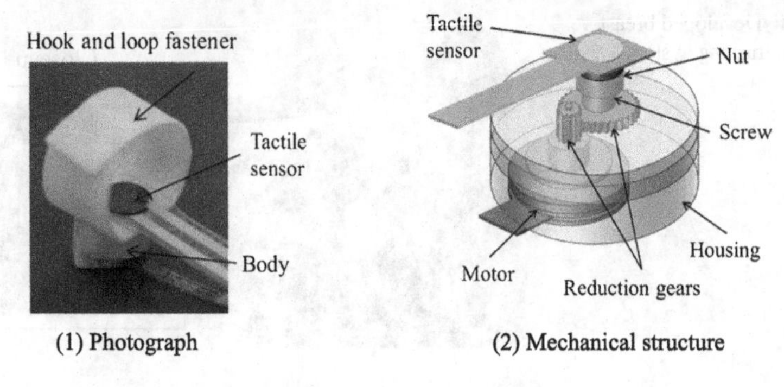

Fig. 5.16 One-dimensional finger pad force display

A haptic device for the finger pads and palm should be small and lightweight, so that it can be attached to the finger pads and palm. The magnitude of the applied force needs to be sufficiently large, so that it can present reality in virtual object manipulation. The device should be easy to wear and should not obstruct the movement of the hand. Previously developed haptic devices for the fingertips or finger pads [54–57] have been problematic in that forces could not be applied to two finger pads of a finger and palm simultaneously because of the size limitation of the device or because the applied forces were insufficient from the low-torque motor. Therefore, we developed a 1D force display device, as shown in Fig. 5.16(1), that can be attached to a finger and palm at three points. This 1D force display device consists of a body, a hook-and-loop fastener (which is wrapped around the finger pad), and a tactile sensor. The mechanical structure of the device is shown in Fig. 5.16(2). The finger pad force display device is driven by a flat-type brushless motor equipped with a Hall sensor (EC10, Maxon Motor). Reduction gears with a reduction ratio of 3.67 are attached to the rotor of the motor, and a screw mechanism with a pitch of 0.5 mm is attached to the output gear axis of the reduction gears. The nut of the screw mechanism moves up and down with the rotation of the motor with a range of movement of 4 mm and contacts the finger pad. The mass and output force are 4.8 g and 2.6 N, respectively. The finger pad force display device contains a tactile sensor (FlexiForce by Nitta Co.). Hence, the contact force is closed-loop controlled.

5.3.3.4 Experiment

We evaluated the effectiveness of the hand haptic interface in a passive touch test in which subjects put their index fingers on the virtual polyhedron with 50 plates consisting of isosceles triangles and trapezoids, as shown in Fig. 5.17. The subjects did not move their index finger, and the polyhedron rotated with about 3 rad/s around an axis that is orthogonal to the flexion/extension axis of the index finger. The force responses by the 1D force display devices at the passive touch of the rotating polyhedron are shown in Fig. 5.18, in which the force display points are (a) the middle

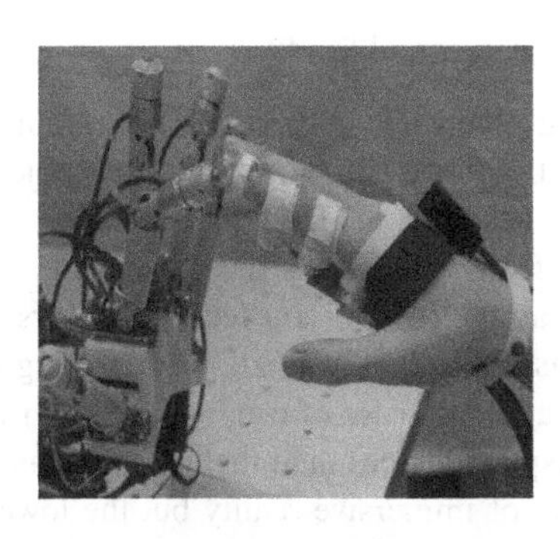

(1) Hand haptic interface

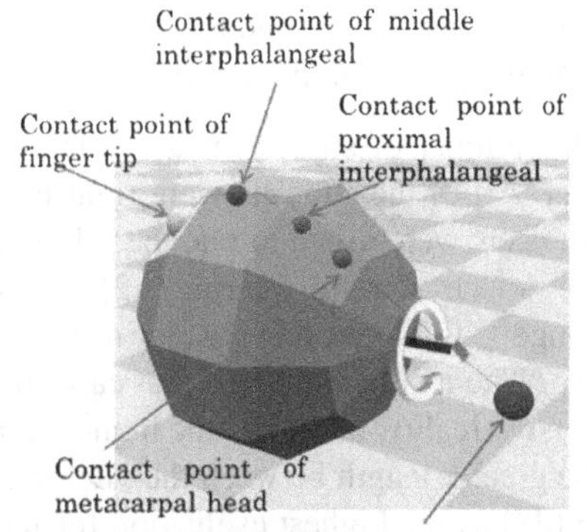

(2) Rotating polyhedron with 50 plates

Fig. 5.17 Experimental setup

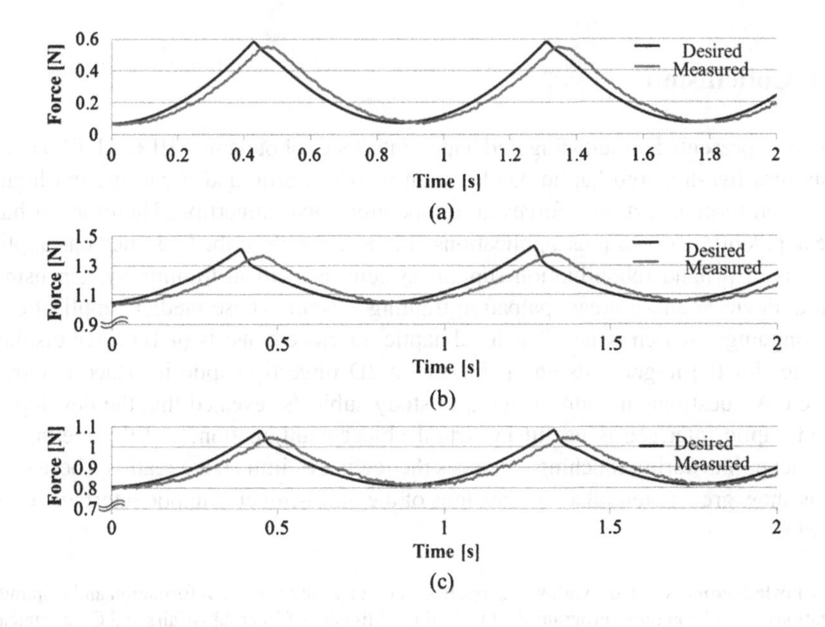

Fig. 5.18 Force responses of 1D force display devices at passive touching of a rotating polyhedron with 50 plates

phalange, (b) the proximal phalange, and (c) the metacarpal head. The responses are almost the same for each contact point, and the maximum force error was 0.12 N.

We also tested the following four cases of force display points by 1D force devices for the index finger:

(a) Middle phalange and proximal phalange
(b) Middle phalange and metacarpal head

(c) Proximal phalange and metacarpal head
(d) Middle phalange, proximal phalange, and metacarpal head

The number of subjects was 10, including nine male subjects and one female subject. All of the subjects were right-handed. After contacting the virtual object, the subjects were asked to rate the following questions: (1) Do you feel the weight of the device you are wearing? (2) Do you feel bothered by the device you are wearing? (3) Do you feel ease of use at operating? (4) Do you feel an immersive reality? The results indicated that case (b) has the lowest evaluation for feeling of immersive reality, but it has the highest evaluation for ease of use, which is caused by the greater length between the 1D force display devices than that of other cases. Case (d) has the highest evaluation for feeling of immersive reality but the lowest evaluation for ease of use. That is, the hand haptic interface enhances the reality in the virtual environment when the user touches a virtual object with his/her fingertips and finger pads.

5.4 Conclusion

We have presented a multi-fingered haptic interface robot named HIRO. HIRO consists of a five-fingered haptic hand and an interface arm, and it can present high-precision three-directional forces at an operator's five fingertips. Therefore, it has great potential for various applications. In fact, we described its medical applications of a hand rehabilitation support system, a medical training system using plural devices, and a breast palpation training system. These medical applications are ongoing research issues. The hand haptic interface consists of 1D force display devices for the finger pads and palm, and a 3D fingertip haptic interface is introduced. A questionnaire administered to study subjects revealed that the developed hand haptic interface is useful in virtual object manipulation, and the feeling of contact points during touching enhances the feeling of immersive reality. These systems show great potential as applications of the multi-fingered haptic interface robot HIRO.

Acknowledgements This work was supported in part by the Strategic Information and Communications R&D Promotion Program (SCOPE) of the Ministry of Internal Affairs and Communications and by a Grant-in-Aid for Scientific Research from JSPS, Japan ((B) No. 23360184).

References

1. Magnenat-Thalmann, N., & Bonanni, U. (2006). Haptics in virtual reality and multimedia. *IEEE Multimedia, 13*(3), 6–11.
2. Najdovski, Z., & Nahavandi, S. (2008). Extending haptic device capability for 3D virtual grasping. In *Haptics: perception, devices and scenarios* (pp. 494–503). Proc. sixth int. conf. EuroHaptics 2008. Berlin: Springer.

3. Kawasaki, H., & Hayashi, T. (1993). Force feedback glove for manipulation of virtual objects. *Journal of Robotics and Mechatronics*, *5*(1), 79–84.
4. Bouzit, M., Burdea, G., Popescu, G., & Boian, R. (2002). The rutgers master II—new design force-feedback glove. *IEEE/ASME Transactions on Mechatronics*, *7*(2), 256–263.
5. Fontana, M., Dettori, D., Salsedo, F., & Bergamasco, M. (2008). Mechanical design of a novel hand exoskeleton for accurate force displaying. In *Proc. of ICRA 2009* (pp. 1074–1709).
6. Website of Immersion Corporation. http://www.immersion.com/3d/products/cyber_force.php.
7. Inaba, G., & Fujita, K. (2006). A pseudo-force-feedback device by fingertip tightening for multi-finger object manipulation. In *Proc. of EuroHaptics 2006*.
8. Ueda, Y., & Maeno, T. (2004). Development of a mouse-shaped haptic device with multiple finger inputs. In *Proc. of IROS 2004* (pp. 2886–2891).
9. Yoshikawa, T., & Nagara, A. (2000). Development and control of touch and force display devices for haptic interface. In *Proc. of SYROCO'00* (pp. 427–432).
10. Adachi, Y., et al. (2002). Development of a haptic device for multi fingers by macro-micro structure. *Journal of the Robotics Society of Japan*, *20*(7), 725–733.
11. Walairacht, S., Ishii, M., Koike, Y., & Sato, M. (2001). Two-handed multi-fingers string-based haptic interface device. *IEICE Transactions on Information and Systems*, *E84D*(3), 365–373.
12. Montoy, M., Oyarzabal, M., Ferre, M., Campos, A., & Barrio, J. (2008). MasterFinger: multi-finger haptic interface for collaborative environment. In *Proc. of EuroHaptics 2008* (pp. 411–419).
13. Kawasaki, H., & Mouri, T. (2007). Design and control of five-fingered haptic interface opposite to human hand. *IEEE Transactions on Robotics*, *23*(5), 909–918.
14. Yokokohji, Y., Muramori, N., Sato, Y., Kimura, T., & Yoshikawa, T. (2004). Design and path planning of an encountered-type haptic display for multiple fingertip contacts based on the observation of human grasping behavior. In *Proc. of ICRA 2004* (pp. 1986–1991).
15. Nakagawara, S., Kajimoto, H., Kawakami, N., Tachi, S., & Kawabuchi, I. (2005). An encounter-type multi-fingered master hand using circuitous joints. In *Proc. of ICRA 2005* (pp. 2667–2672).
16. Dovat, L., Lambercy, O., Gassert, R., Maeder, T., Milner, T., Leong, T. C., & Burdet, E. (2008). HandCARE: a cable-actuated rehabilitation system to train hand function after stroke. *IEEE Transactions on Neural Systems and Rehabilitation Engineering*, *16*(6), 582–591.
17. Connelly, L., Jia, Y., Toro, M., Stoykov, M. E., Kenyon, R., & Kamper, D. (2010). A pneumatic glove and immersive virtual reality environment for hand rehabilitation training after stroke. *IEEE Transactions on Neural Systems and Rehabilitation Engineering*, *18*(5), 551–559.
18. Okamura, A. M., Webster, R. J. III, Nolin, J. T., Johonson, K. W., & Jafry, H. (2003). The haptic scissors: cutting in virtual environments. In *Proc. of ICRA 2003* (pp. 828–833).
19. Rosenberg, L. B., & Stredney, D. (1996). A haptic interface for virtual simulation of endoscopic surgery. In H. Sieburg et al. (Eds.), *Health care in the information age*. Amsterdam: IOS Press/Ohmsha.
20. Langrana, N., Burdea, G., Ladeji, J., & Dinsmore, M. (1997). Human performance using virtual reality tumor palpation simulation. *Computers & Graphics*, *21*(4), 451–458.
21. Howe, R. D., Peine, W. J., Kontarinis, D. A., & Son, J. S. (1995). Remote palpation technology. *IEEE Engineering in Medicine and Biology Magazine*, 318–323.
22. Endo, T., Kawasaki, H., Mouri, T., Ishigure, Y., Shimomura, H., Matsumura, M., & Koketsu, K. (2011). Five-fingered haptic interface robot: HIRO III. *IEEE Transactions on Haptics*, *4*(1), 14–27.
23. Hioki, M., Kawasaki, H., Sakaeda, H., Nishimoto, Y., & Mouri, T. (2011). Finger rehabilitation support system using a multifingered haptic interface controlled by a surface electromyogram. *Journal of Robotics*, *2011*, 10.
24. Kawasaki, H., Mouri, T., & Ikenohata, S. (2007). Multi-fingered haptic interface robot handling plural tool devices. In *Proc. of world haptics 2007* (pp. 397–402).

25. Endo, T., Tanimura, S., & Kawasaki, H. (2011). Development of a surgical knife device for a multi-fingered haptic interface robot. In *Preprints of the 18th IFAC world congress* (pp. 6460–6465).
26. Endo, T., Tanimura, S., & Kawasaki, H. (2011). Development of a tweezers-type device for a multi-fingered haptic interface robot. In *Proc. of 2011 IEEE/SICE international symposium on system integration (SII2011)* (pp. 1006–1011).
27. Daniulaitis, V., Alhalabi, M. O., Kawasaki, H., Tanaka, Y., & Hori, T. (2004). Medical palpation of deformable tissue using physics-based model for haptic interface RObot (HIRO). In *Proc. of IROS 2004* (pp. 3907–3911).
28. Alharabi, M. O., Daniulaitis, V., Kawasaki, H., & Hori, T. (2005). Medical training simulation for palpation of subsurface tumor using HIRO. In *Proc. world haptics 2005* (pp. 623–624).
29. Kawasaki, H., Koide, S., Mouri, T., & Endo, T. (2010). Finger pad force display for hand haptic interface. In *Proc. of 6th IEEE conference on automation science and engineering (CASE 2010)* (pp. 374–379).
30. Kawasaki, H., Koide, S., Endo, T., & Mouri, T. (2012). Development of a hand haptic interface and its basic experimental evaluation. In *Proc. of international symposium on innovations in intelligent systems and applications (INISTA 2012)* (5 pp.).
31. Website of UAB Aksonas. http://robothand.eu/en/products/robotic_hands/.
32. Website of Marutomi Seiko Co., Ltd. http://www.maru-tomi.co.jp/english/index_e.html.
33. Endo, T., Kawachi, Y., Kawasaki, H., & Mouri, T. (2008). FPGA-based control for the wire-saving of five-fingered haptic interface. In M. Ferre (Ed.), *Haptics: perception, devices and scenarios* (pp. 536–542). Proc. sixth int'l conf. EuroHaptics 2008. Berlin: Springer.
34. Carignan, C., & Liszka, M. (2005). Design of an arm exoskeleton with scapula motion for shoulder rehabilitation. In *Proc. of ICRA 2005* (pp. 524–531).
35. Gupta, A., & O'Malley, M. K. (2006). Design of a haptic arm exoskeleton for training and rehabilitation. *IEEE/ASME Transactions on Mechatronics*, *11*(3), 280–289.
36. Mahoney, R. M., van der Loos, H. F. M., Lum, P. S., & Burgar, C. (2003). Robotic stroke therapy assistant. *Robotica*, *21*, 33–44.
37. Oblak, J., Cikajlo, I., & Matjacic, Z. (2010). Universal haptic drive: a robot for arm and wrist rehabilitation. *IEEE Transactions on Neural Systems and Rehabilitation Engineering*, *18*(3), 239–302.
38. Kawasaki, H., Ito, S., Ishigure, Y., Nishimoto, Y., Aoki, T., Mouri, T., Sakaeda, H., & Abe, M. (2007). Development of hand motion assist robot for rehabilitation therapy by patient self-motion control. In *Proc. of IEEE ICORR 2007* (pp. 234–240).
39. Ueki, S., Kawasaki, H., Ito, S., Nishimoto, Y., Abe, M., Aoki, T., Ishigure, Y., Ojika, T., & Mouri, T. (2011). Development of a hand-assist robot with multi-degrees-of-freedom for rehabilitation therapy. *IEEE/ASME Transactions on Mechatronics*, *17*(1), 136–146.
40. Freeman, C., Burridge, J., Chappell, P., Lewin, P., & Rogers, E. (2009). Upper limb rehabilitation of stroke participants using electrical stimulation: changes in tracking and EMG timing. In *Proc. of IEEE ICORR 2009* (pp. 289–294).
41. Hu, X. L., Tong, K. Y., Song, R., Zheng, X. J., & Leung, W. W. F. (2009). A randomized controlled trial on the recovery process of wrist rehabilitation assisted by electromyography (EMG)-driven robot for chronic stroke. In *Proc. of IEEE ICORR 2009* (pp. 289–294).
42. Hioki, M., & Kawasaki, H. (2009). Estimation of finger joint angles from sEMG using a recurrent neural network with time-delayed input vectors. In *Proc. of IEEE ICORR 2009* (pp. 289–294).
43. Bu, N., Okamoto, M., & Tsuji, T. (2009). A hybrid motion classification approach for EMG-based human-robot interface using Bayesian and neural networks. *IEEE Transactions on Robotics*, *25*(3), 502–511.
44. Oskoei, M. A., & Hu, H. (2007). Myoelectric control system—a survey. *Biomedical Signal Processing and Control*, *2*, 275–294.
45. Honma, S., & Wakamatsu, H. (2004). Cutting moment analysis of materials by the saw for force display system. *Transactions of the Virtual Reality Society of Japan*, *9*(3), 319–326 (in Japanese).

46. Webster, R. J. III, Memisevic, J., & Okamura, A. M. (2005). Design considerations for robotic needle steering. In *Proc. of the 2005 IEEE international conference on robotics and automation* (pp. 3599–3605).
47. Tzafestas, C. S., Christopoulos, D., & Birbas, K. (2006). Haptic display improves training and skill assessment performance in a virtual paracentesis simulator: a pilot evaluation study. In *Proc. of euro haptics*.
48. Ota, D., Loftin, B., Saito, T., Lea, R., & Keller, J. (1995). Virtual reality in surgical education. *Computers in Biology and Medicine*, *25*(2), 127–137.
49. Burdea, G. (1996). *Force and touch feedback for virtual reality*. New York: Wiley.
50. Langrana, N., Burdea, G., Ladeji, J., & Dinsmore, M. (1997). Human performance using virtual reality tumor palpation simulation. *Computers & Graphics*, *21*(4), 451–458.
51. Cotin, S., Delingette, H., & Ayache, N. (1999). Real-time elastic deformations of soft tissues for surgery simulation. *IEEE Transactions on Visualization and Computer Graphics*, *5*(1), 62–73.
52. Kobayashi, M., Endo, T., Goto, T., & Kawasaki, H. (2010). Transmission of force and position information using a hand image and a multi-fingered haptic device. In *Proc. of SI2010* (pp. 190–191) (in Japanese).
53. Johanson, R. S., & Vallbo, A. B. (1979). Tactile sensibility in the human hand: relative and absolute densities of four types of mechanoreceptive units in glabrous skin. *Journal of Physiology*, *286*, 283–300.
54. Grieve, T., Sun, Y., Hollerbach, J. M., & Mascaro, S. A. (2009). 3-d force control on the human fingerpad using a magnetic levitation device for fingernail imaging calibration. In *Proc. of 3rd joint eurohaptics conference and symposium on haptic interfaces for virtual environment and teleoperator system* (pp. 411–416).
55. Minamizawa, K., Fukamachi, S., Kajimoto, H., Kawakami, N., & Tachi, S. (2008). Wearable haptic display to present mass and internal dynamics of virtual objects. *Transactions of the Virtual Reality Society of Japan*, *13*(1), 15–24.
56. Aoki, T., Mitake, H., Hasegawa, S., & Sato, M. (2009). Wearable haptic device to present contact sensation based on cutaneous sensation using thin wires. *Transactions of the Virtual Reality Society of Japan*, *14*(3), 421–428.
57. Koo, I. M., Jung, K., Koo, J. C., Nam, J., & Lee, Y. K. (2008). Development of soft-actuator based wearable tactile display. *IEEE Transactions on Robotics*, *24*(3), 548–558.

Chapter 6
Two-Handed Multi-finger String-Based Haptic Interface SPIDAR-8

Katsuhito Akahane, Jungwoo Hyun, Itsuo Kumazawa, and Makoto Sato

Abstract In this chapter, the authors propose a system for implementing direct, two-handed, multi-finger manipulations in a virtual world. To implement an easier-to-use means of direct, two-handed, multi-finger manipulations, it is necessary to give a sense of touch to the hands to make them feel like they are grasping an object and to make it easier for the user to understand the positional relationship between the user's hands and the virtual object. Therefore, the authors constructed a system that provides force feedback to eight fingertips and that can display virtual hands (Walairacht et al. in Proceedings of the ninth international conference on artificial reality and tele-existence, pp. 233–236, 1999) making the same movements as the user's actual hands in real time. Therefore, the authors constructed a system that provides force feedback to eight fingertips (called SPIDAR-8 Walairacht et al. in Proceedings of the IEEE international symposium on intelligent signal processing and communication systems, pp. 569–572, 1999 and Proceedings of the IEEE virtual reality 2000, p. 293, 2000). SPIDAR-8 is a device that can measure the positions of eight points and provide force feedback by using strings. Force feedback can be provided to a fingertip by attaching a cap-like apparatus to the ends of the strings and fitting this apparatus onto the fingertip. The virtual hands are displayed according to the eight fingertip positions calculated by SPIDAR-8. Then hand shapes are estimated from the positions of the eight points, and these hands are displayed together with the virtual object via computer graphics. Virtual hands making the same movements as the user's actual hands are displayed in this way. With this kind of system, the user can integrate the haptic and visual information that is presented to efficiently perform direct, two-handed, multi-finger manipulations.

K. Akahane (✉) · J. Hyun · I. Kumazawa · M. Sato
Tokyo Institute of Technology, 4259 R2-13 Nagatsuta, Midori, Yokohama 226-8503, Japan
e-mail: kakahane@hi.pi.titech.ac.jp

J. Hyun
e-mail: rurouni75@gmail.com

I. Kumazawa
e-mail: kumazawa@isl.titech.ac.jp

M. Sato
e-mail: msato@pi.titech.ac.jp

I. Galiana, M. Ferre (eds.), *Multi-finger Haptic Interaction*,
Springer Series on Touch and Haptic Systems, DOI 10.1007/978-1-4471-5204-0_6,

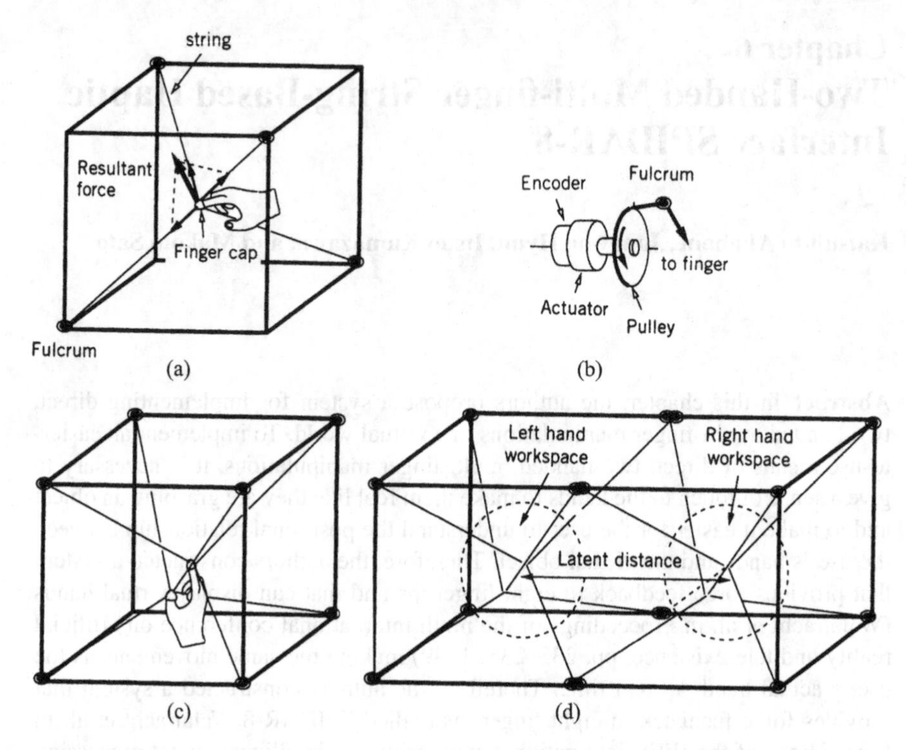

Fig. 6.1 SPIDAR systems: (**a**) SPIDAR-I; (**b**) Motor, encoder, pulley, and string; (**c**) SPIDAR-II; (**d**) Both-Hands-SPIDAR

6.1 Introduction

In 1992, Sato and Hirata had developed another type of string-based haptic interface device named SPIDAR [1]. The SPIDAR is abbreviated from the SPace Interface Device for Artificial Reality. The initial prototype of the device has 4 strings attached to a finger cap and allow a user to use a fingertip to interact with the virtual environment. The length of each string is measured by the rotary encoder and the force is generated by controlling the tension of strings.

In the later year, the SPIDAR system had been modified by Ishii to allow the use two fingers, thumb and index finger. The interface to each finger requires 4 strings, thus the system now uses 8 strings in total. Various types of SPIDARS are illustrated in Fig. 6.1. SPIDAR-II [2] makes possible for grasping a virtual object by both fingers with force feedback.

The further implementation of SPIDAR-II is the Both-Hands SPIDAR [3] by Ishii. Two system of SPIDAR-II are combined on a single support frame. Each hand of the user has separate spherical working space, which is about 30 cm in diameter. In the manipulation space, the user's hands are separated by the latent distance but the corresponding virtual hands are touching in the virtual scene.

Force display is mainly realized by complicated structure. These complicated and heavy structures both make the user feel uncomfortable and limited his moving

Fig. 6.2 BIG-SPIDAR

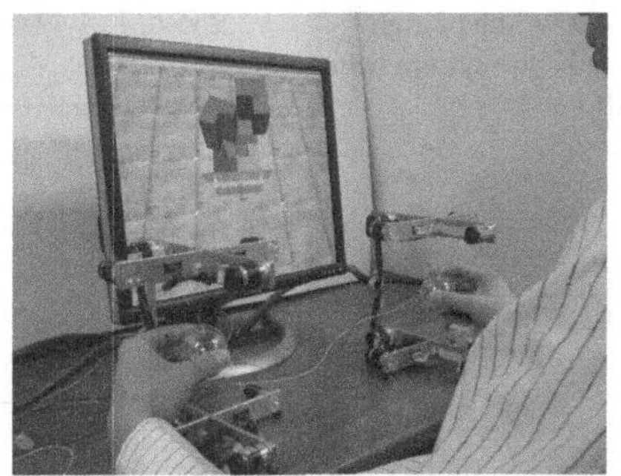

Fig. 6.3 SPIDAR-G

freedom. We made a string-motor structure device, called BIG-SPIDAR [4] and shown in Fig. 6.2, to realize the position measurement and force display in a human-scale virtual environment. This is a kind of simple mechanism, light weight and moving free system. A system is developed which allow the user not only move around but also feeling force sensation in it.

Force feedback for the device is tension based and is characterized by 7 degrees of freedom (DOF); 3 DOF for translation, 3 DOF for rotation, and 1 DOF for grasp. The SPIDAR-G (Space Interface Device for Artificial Reality with Grip) [5] is shown in Fig. 6.3 will allow users to interact with virtual objects naturally by manipulating two hemispherical grips located in the center of the device frame.

We developed a new haptic interface SPIDARG&G [6] system shown in Fig. 6.3 to take advantage of the user's existing two hands skills for completing the continuous and compound task in a virtual world. A user cannot only manipulate the 3D objects with 6DOF translational and rotational motion by using the grip which located in the workspace of two frames, but also get the 6DOF force feedback, 3DOF for translation force, 3DOF for rotation force. SPIDAR-G&G system can offer direct two-handed 6DOF manipulation for virtual environment via the sense of touch.

The further implementation of SPIDAR-G is an inner string haptic device SPIDAR-I [7]. This device is designed analytically to improve performance of calculation. Therefore, this device can calculate position and orientation more faith-

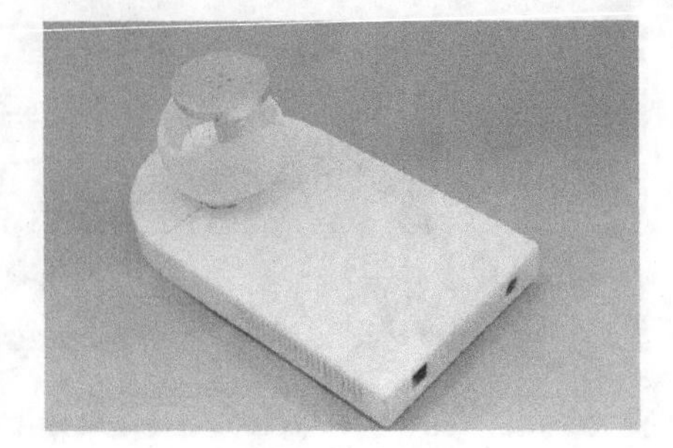

Fig. 6.4 SPIDAR-I

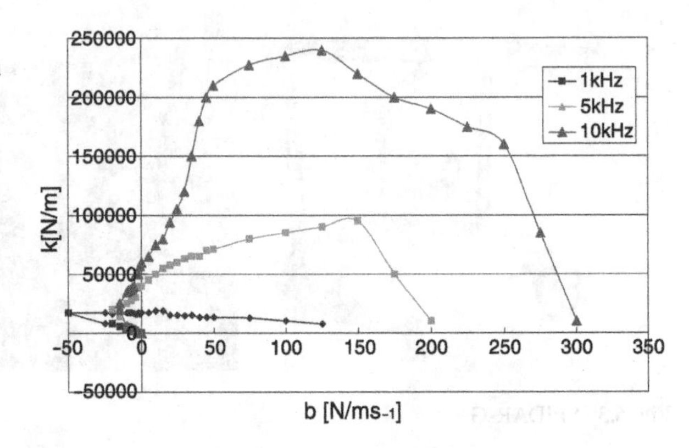

Fig. 6.5 Z-Width of high definition haptic system with SPIDAR

fully. This device has structure that strings are inside a grip,and it became possible to miniaturize by placing mechanics inside a grip too. This device can measure posture of 6 degrees of freedom (3 for translation, 3 for rotation) and display force of 6 degree of freedom (Fig. 6.4).

Force feedback bandwidth is updated at 10 kHz [8, 9], and the graphics refresh rate is 60 Hz (video rate). We developed a 10 kHz high definition haptic rendering system for general VR applications. We could achieve both stability and fidelity of haptic rendering with SPIDAR family (Fig. 6.5).

6.2 System Design and the Construction of SPIDAR-8

In this section, we give explanation about design and construction of the proposed haptic interface device. The content of this section starts with the description of general specification of developing a hand haptic interface device. Then we set up our design policy from the specification and requirements. The problem concerning the design of the device was discussed and we have presented the compromising solution to solve the design problem. At the end of this section, we describe hardware construction of the propose device.

6.2.1 Design Specification

General specification for the development of hand haptic interface device for objects manipulation in the virtual environment can be pointed out as follows.

- Precision—The device should have sufficient position sensing accuracy and adequate force display ability.
- Light in weight—Weight of the device should not easily lead to user fatigue and reduce the portability.
- Compact and simple structure—Simple structure device, in general, is easy to control and maintain. The structure of the device should not limit natural motion and usability which are compatible with the dimension and motion range of the hands and fingers.
- Safe—The device should not be harmful, which could cause any injury to the user.
- Inexpensive—The cost of the device and also the cost of maintenance should not be too expensive. This implies to the cost of computation as well. Inexpensive device could easily gain the popularity and widespread use.

6.2.2 Design Policy

We have set up the design policy of the proposed haptic interface device based on general specification described above and the requirements (Fig. 6.6). The device is being built as a string-based haptic interface device in which this type of device has shown many advantages. Moreover, the device is designed to provide the user with direct manipulation environment when performing virtual objects manipulation task.

- *Interface device for both hands and multi fingers*. Two-handed interface device is introduced as a potential means to improve the ability in virtual objects manipulation task performed by hands. Providing both hands with one single working space, the user is allowed to use both hands in cooperative way to manipulate the objects. In addition, the perception of various physical properties, such as shape or surface smoothness, of the manipulated object can be easily perceived by the help of both hands.

  As object manipulation performed by a hand, it concerns with various ways of grasping such as a power grasp by the whole hand or dextrous grasp by only the fingertips. In most cases of dextrous manipulation of object, the more number of finger used results in more stable of the grasping and the improvement of efficiency in the manipulation. Thus, to simulate the dextrous manipulation of virtual objects, the use of multi-fingered interface device is advisable.
- *Sensing positions and displaying force feedback by means of strings*. Haptic interface device that senses positions on the user's hands by means of strings and uses the tension of strings for displaying force has own many advantages. String-based haptic interface device is usually simple, easy to control, and inexpensive.

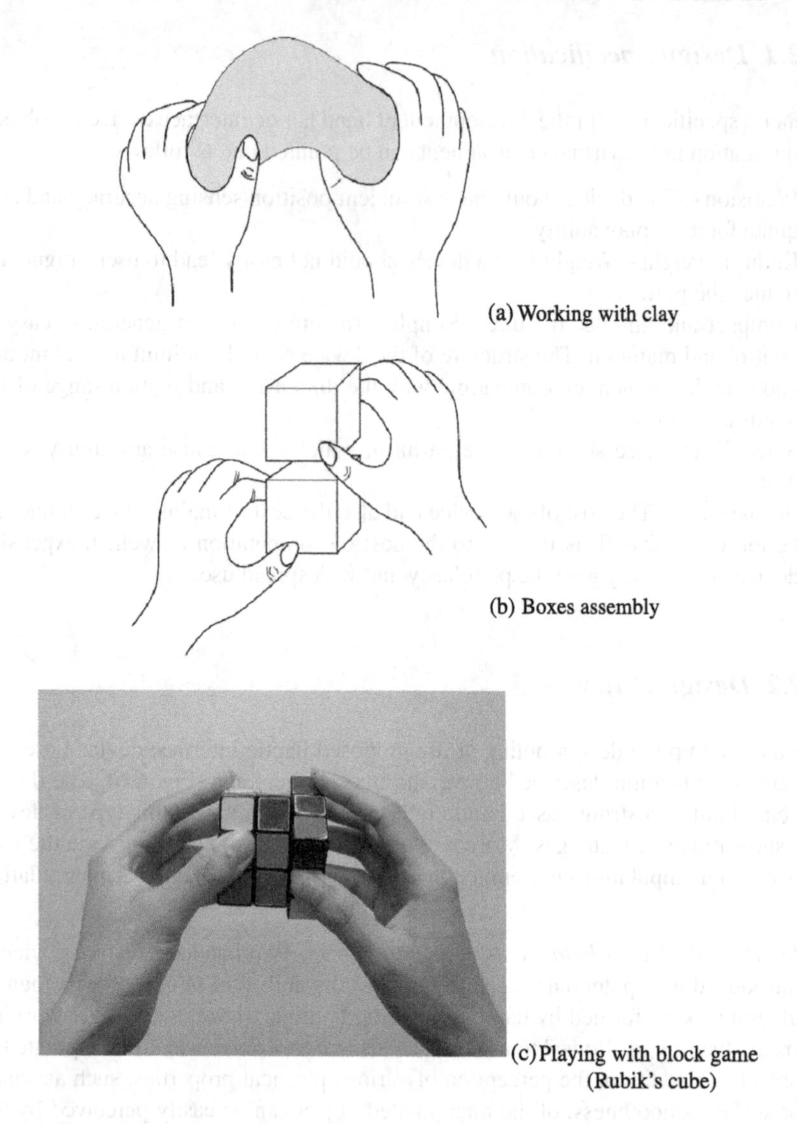

(a) Working with clay
(b) Boxes assembly
(c) Playing with block game (Rubik's cube)

Fig. 6.6 Examples of task requiring both hands

The device is very light in weight and safe as the actuators are placed remotely from the user's hands. However, to be able to control more degrees of freedom on the user's hands, the number of string used must be increased in the proportional ratio. Therefore, the problem is the interference of string and this problem must be carefully taken into consideration.

- *Providing direct manipulation environment.* String-based haptic interface device according to our design attaching the strings directly to the user's fingers by means of small light-weighted finger caps. This minimized attachment allows the user to have free movement of the hands and fingers in the manipulation space and be able to directly manipulate the virtual objects by hands. Unlike the other types of device where the hands are heavily attached with complex mechanical structures or being hidden inside the instrument glove, the user often has the feeling of operating some kind of machinery tool.

6.2.3 Design Problem and Solution

From our design policy described in the previous subsection, the use of strings has raised up some design problems. The problems are mainly concerned with force feedback ability and the interference of string. We take into consideration the trade-off between the use of strings and the ability of displaying force to introduce the solutions for solving the problems.

- Problems concerning the use of strings
  - Number of strings—As stated before, string-based haptic interface devices use strings for displaying force feedback. To be able to control more degrees of freedom of force, it is necessary to use large number of strings. Kawamura and Ito had made an investigation and finally concluded that $n + 1$ strings were necessary and sufficient to make force and moment with n degrees of freedom [10]. Therefore, to be able to display force feedback on a finger in any direction in 3D space, 4 strings are necessary.
  - String interference—In the case of a multi-fingered system and the actuators are placed remotely on the device frame, if we attach 4 strings to each finger, there will be too many strings from the actuators attaching to the user's hands. In this way, it limits the movement of the user's hands and fingers and the strings can easily interfere with each other and also interfere with the user's hands.
- Solutions for the design of the proposed device
  - Reducing number of strings—By measuring the length of 3 strings and knowing 3 positions of string's fulcrums at each actuator, it is enough to compute a position of a finger in 3D space. However, the use 3 strings cannot display force feedback on a finger to every direction in 3D. This problem is solved by suitable arrangement of strings from the actuators attached to each finger of the user and force feedback can be displayed in most direction in 3D. The limitation is force feedback cannot be displayed only to the direction that all strings cannot produce any tension.
  - Arranging actuators—The method of reducing number of strings used can also reduce the interference of strings. However, the arrangement of the actuators attached on the device frame should be carefully considered. The way that 3 strings from 3 actuators attaching to a finger should be able to produce sufficient force display ability with minimum string interference.

To compromise between force display ability and string interference of the proposed device, three strings from three actuators placed remotely on the device frame are attached to each finger for sensing fingertip position and displaying force feedback. For each finger, each of the actuators is placed on the frame along x-, y, and z-axes. Three strings connected to a finger and three positions of the actuators establish a triangular-base prism of force display volume and we, hereby, call it as force cone. If a force vector resulted from the collision of a finger to the virtual object is inside the force cone, force feedback can be generated using the tension of 3 strings. In the case that force vector lies outside but is possible to project back to the force cone, the appropriate force feedback can be displayed using the projected tension of strings. In the worst case, if force vector is in the direction that cannot produce any tension of string, the other fingers on the same hand or the fingers on anther side of the hand still can realize force feedback.

6.2.4 System Design

6.2.4.1 Working Space of Both Hands

Corlett defined the average work range of left and right hands for a male adult [11] as shown in Fig. 6.7. From this picture, it can be seen that the movement of the arms define two spherical working space about 60 cm in radius partially overlapping each other. By assuming that the user only performs a task in front of him or her and without turning his or her body in other directions, we can approximately define the working space of both hands for our proposed device by the overlapped space in the front of the user, which is approximated as a spherical workspace about 30 cm in radius. The structure of the proposed device is designed to be a rectangular cubic frame. The frame is 80 cm in width, 60 cm in hight, and 60 cm in depth. The space enclosed by the frame is large enough for the working space of both hands as defined above.

6.2.4.2 The Actuator

An actuator of the proposed device consists of a DC motor, a rotary encoder, a pulley, and a string. Figure 6.8 shows the photograph of a pair of the actuators. The rotary encoder is connected at the back of the motor where on the same shaft a pulley is attached in front. The light-weighted, non-elastic polyethylene string is wound inside the pulley. The actuator is attached to the frame by means of the supporting device where the string is passing through the string fulcrum. While the string is pulled by the finger or is wound back by the motor, the rotary encoder generates the signal pulse. The signal pulse is counted by the counter, then the length of the string can be found. To generate the tension on the string, the amount of electrical current entering the motor is controlled from the computer. Force feedback is composed from the tension of the strings and displayed to the user.

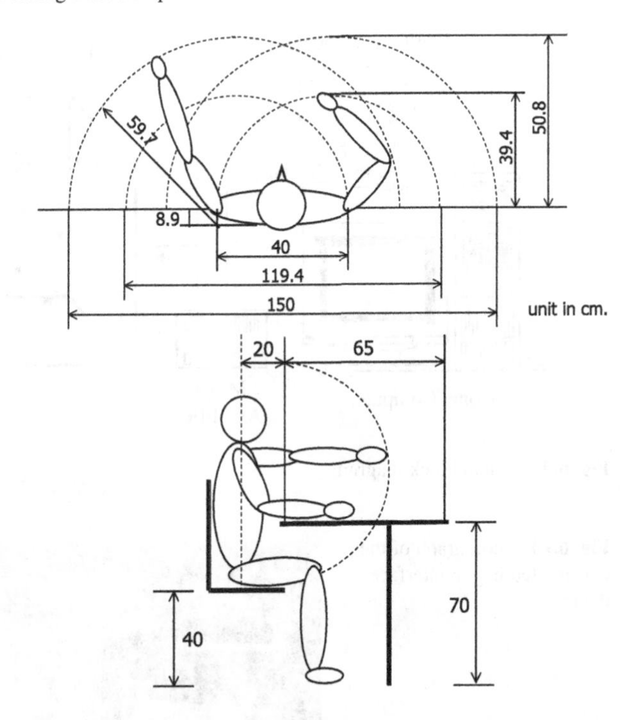

Fig. 6.7 Work ranges of hands

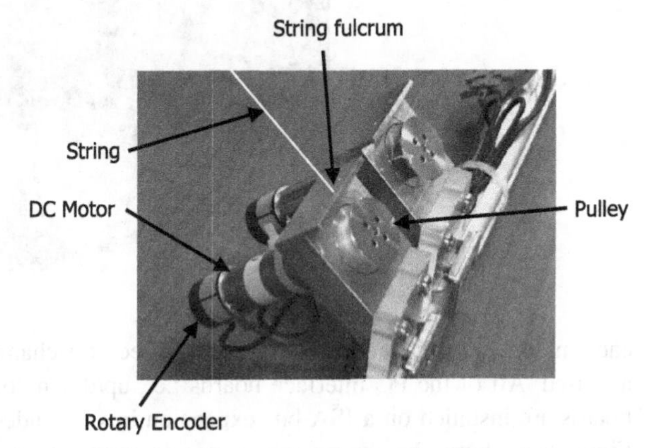

Fig. 6.8 Actuator

6.2.5 System Construction

6.2.5.1 System Block Diagram

System block diagram is shown in Fig. 6.9. A personal computer is used for controlling the system. This computer is referred as haptic server. The signal pulses read from each rotary encoder are counted by the up/down counter boards. Since there are 24 actuators on the system, therefore, six of 4-channel up/down counter boards are used. The haptic server controls the amount of current through motor amplifier to

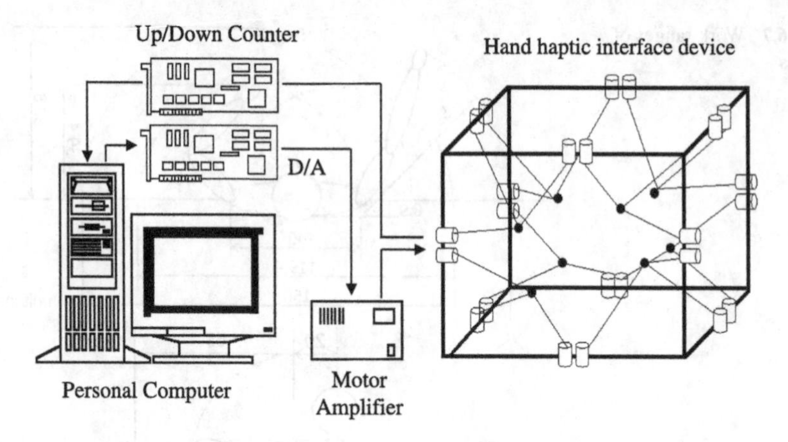

Fig. 6.9 System block diagram

Fig. 6.10 Photograph of the constructed haptic interface device

each motor by the D/A converter boards. Three of 8-channel D/A converter boards are used. All of the PC interface boards i.e., up/down counter and D/A converter boards are installed on a ISA-bus expansion box extended from the haptic server. Figure 6.10 shows the photograph of the constructed haptic interface device.

6.2.5.2 Device Frame

There are 24 actuators attached on the frame as shown in Fig. 6.11. Each actuator is labelled with the sensing finger in the figure. Three actuators each of them is on x-, y-, and z-axis of the frame are interfaced with one finger. The locations of each actuator with the corresponding sensing finger are summarized as shown in Table 6.1.

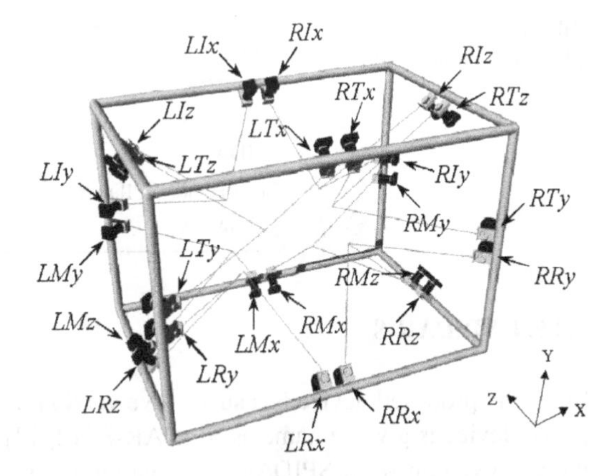

Fig. 6.11 Locations of twenty-four actuators on the frame

Table 6.1 The actuators to the corresponding fingertip

Actuator	Sensing finger
(LT_x, LT_y, LT_z)	Left Thumb
(LI_x, LI_y, LI_z)	Left Index
(LM_x, LM_y, LM_z)	Left Middle
(LR_x, LR_y, LR_z)	Left Ring
(RT_x, RT_y, RT_z)	Right Thumb
(RI_x, RI_y, RI_z)	Right Index
(RM_x, RM_y, RM_z)	Right Middle
(RR_x, RR_y, RR_z)	Right Ring

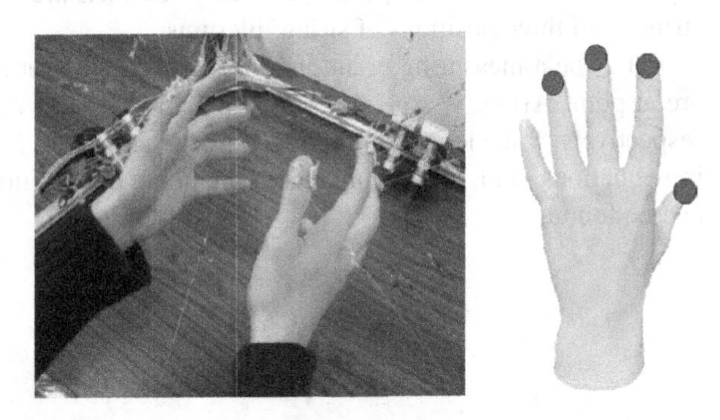

Fig. 6.12 Fingertips attachment

6.2.5.3 Fingertip Attachment

Figure 6.12 shows that thumb, index finger, middle finger, and ring finger on left and right hands of the user are wearing small cap-liked devices on the fingertips. Each finger cap is attached by three strings from three different actuators as already explained.

Table 6.2 Comparison of different version of SPIDAR systems

Name	No. of finger	No. of string
SPIDAR-I	1	4
SPIDAR-II	2	8
Both-Hands SPIDAR	4	16
SPIDAR-8	8	24

6.2.5.4 SPIDAR-8

Since the proposed device is a successive version of the SPIDAR systems, the proposed device is given a name as SPIDAR-8 [12, 13]. The numerical digit followed the abbreviation word SPIDAR is indicating the number of interfaced fingers. We hereby refer to the proposed device by its name. Table 6.2 summarizes the number of strings used for different version of SPIDAR systems.

6.3 Position Sensing and Force Feedback of SPIDAR-8

6.3.1 Position Sensing

SPIDAR-8 senses each fingertip position on the user's hand using the length of three strings. To measure the length of each string correctly, each motor is controlled to pull the string by small amount of force about 0.3 N to straighten the string. Figure 6.13(a) show a fingertip position to be measured using the length of three strings and three positions of string fulcrums.

Let P be a measuring point on a fingertip of the user and three string fulcrums are at point $A_1(x_1, y_1, z_1)$, $A_2(x_2, y_2, z_2)$, and $A_3(x_3, y_3, z_3)$ on x-, y-, and z-axis respectively. If the length of three strings l_i where $i = 1, 2, 3$ are known, the following equations hold. By solving these equations, the position of point $P(p_x, p_y, p_z)$ can be found.

$$l_1^2 = \|P - A_1\|^2 \tag{6.1}$$

$$l_2^2 = \|P - A_2\|^2 \tag{6.2}$$

$$l_3^2 = \|P - A_3\|^2 \tag{6.3}$$

We have elaborated the computation of point P by constructing vector components as shown in Fig. 6.13(b). Vector $\mathbf{n}_1$ and vector $\mathbf{n}_2$ are unit vectors along vector $\overrightarrow{A_2A_1}$ and vector $\overrightarrow{A_3A_1}$ respectively. An unit vector $\mathbf{n}_3$ is the cross product of $\mathbf{n}_1$ and $\mathbf{n}_2$. An angle between vector $\overrightarrow{A_2A_1}$ and vector $\overrightarrow{A_3A_1}$ is represented by a parameter θ. We can express $\mathbf{n}_1$, $\mathbf{n}_2$, and $\mathbf{n}_3$ by the following equations.

$$\mathbf{n}_1 = \frac{A_2 - A_1}{\|A_2 - A_1\|} \tag{6.4}$$

Fig. 6.13 Position measurement

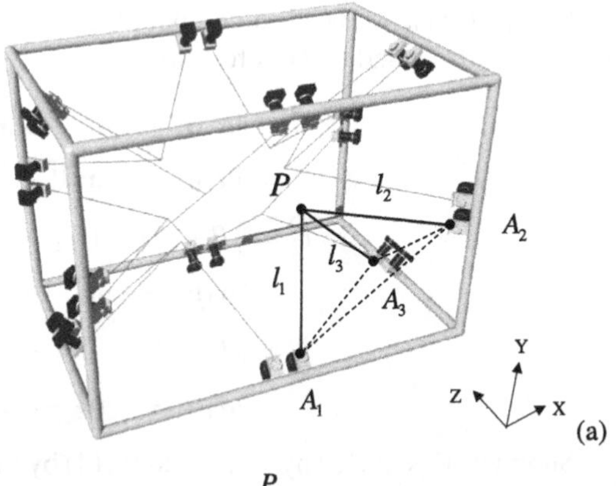

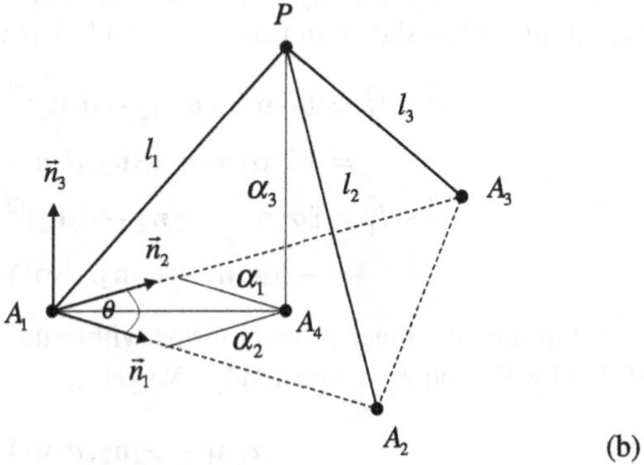

$$\mathbf{n}_2 = \frac{A_3 - A_1}{\|A_3 - A_1\|} \tag{6.5}$$

$$\mathbf{n}_3 = \mathbf{n}_1 \times \mathbf{n}_2 \tag{6.6}$$

Next, d_1 is a distance measured from point A_2 to point A_1 and d_2 is a distance measured from point A_3 to point A_1. Both distances are known as the magnitude of vector $\overrightarrow{A_2A_1}$ and vector $\overrightarrow{A_3A_1}$ respectively.

$$d_1 = \|A_2 - A_1\| \tag{6.7}$$

$$d_2 = \|A_3 - A_1\| \tag{6.8}$$

From Eqs. (6.4), (6.5) and Eqs. (6.7), (6.8) above, we can have

$$A_2 - A_1 = d_1 \cdot \mathbf{n}_1 \tag{6.9}$$

$$A_3 - A_1 = d_2 \cdot \mathbf{n}_2. \tag{6.10}$$

A point $A_4(x_4, y_4, z_4)$ is the orthogonal projection of point P onto a triangular plane $A_1A_2A_3$. The orthogonal distance is measured as α_3. Because vector $\overrightarrow{A_4A_1}$ is the summation of $\alpha_1\mathbf{n}_1$ and $\alpha_2\mathbf{n}_2$, therefore point P can be found by the following equation.

$$P = A_1 + \alpha_1\mathbf{n}_1 + \alpha_2\mathbf{n}_2 + \alpha_3\mathbf{n}_3, \quad \alpha_3 \geq 0 \tag{6.11}$$

In Eq. (6.11), we need to compute the value of α_1, α_2 and α_3. So, we rewrite Eqs. (6.1), (6.2), and (6.3) as follows.

$$\begin{aligned} l_1^2 &= \|\alpha_1\mathbf{n}_1 + \alpha_2\mathbf{n}_2 + \alpha_3\mathbf{n}_3\|^2 \\ &= \|\alpha_1\mathbf{n}_1 + \alpha_2\mathbf{n}_2\|^2 + \alpha_3^2 \end{aligned} \tag{6.12}$$

$$\begin{aligned} l_2^2 &= \|P - A_1 - (A_2 - A_1)\|^2 \\ &= \|\alpha_1\mathbf{n}_1 + \alpha_2\mathbf{n}_2 - d_1\mathbf{n}_1\|^2 + \alpha_3^2 \end{aligned} \tag{6.13}$$

$$\begin{aligned} l_3^2 &= \|P - A_1 - (A_3 - A_1)\|^2 \\ &= \|\alpha_1\mathbf{n}_1 + \alpha_2\mathbf{n}_2 - d_2\mathbf{n}_2\|^2 + \alpha_3^2 \end{aligned} \tag{6.14}$$

Subtracts Eqs. (6.13) by (6.12) and (6.14) by Eq. (6.12), the term α_3 in both cases are eliminated as shown in Eqs. (6.15) and (6.16) respectively.

$$\begin{aligned} l_2^2 - l_1^2 &= \|\alpha_1\mathbf{n}_1 + \alpha_2\mathbf{n}_2 - d_1\mathbf{n}_1\|^2 - \|\alpha_1\mathbf{n}_1 + \alpha_2\mathbf{n}_2\|^2 \\ &= -2(\alpha_1\mathbf{n}_1 + \alpha_2\mathbf{n}_2, d_1\mathbf{n}_1) + d_1^2 \end{aligned} \tag{6.15}$$

$$\begin{aligned} l_3^2 - l_1^2 &= \|\alpha_1\mathbf{n}_1 + \alpha_2\mathbf{n}_2 - d_2\mathbf{n}_2\|^2 - \|\alpha_1\mathbf{n}_1 + \alpha_2\mathbf{n}_2\|^2 \\ &= -2(\alpha_1\mathbf{n}_1 + \alpha_2\mathbf{n}_2, d_2\mathbf{n}_2) + d_2^2 \end{aligned} \tag{6.16}$$

Simplifies the inner product terms within the right hand side of Eqs. (6.15) and (6.16) by K_1 and K_2 respectively. We get

$$(\alpha_1\mathbf{n}_1 + \alpha_2\mathbf{n}_2, d_1\mathbf{n}_1) = K_1 \tag{6.17}$$

$$(\alpha_1\mathbf{n}_1 + \alpha_2\mathbf{n}_2, d_2\mathbf{n}_2) = K_2. \tag{6.18}$$

We can rewrite Eqs. (6.15) and (6.16) by K_1 and K_2 as

$$K_1 = \frac{1}{2}\{d_1^2 - (l_2^2 - l_1^2)\} \tag{6.19}$$

$$K_2 = \frac{1}{2}\{d_2^2 - (l_3^2 - l_1^2)\}. \tag{6.20}$$

From Eqs. (6.17) and (6.18), we can get

$$\alpha_1 + \cos\theta \cdot \alpha_2 = \frac{K_1}{d_1} \tag{6.21}$$

$$\cos\theta \cdot \alpha_1 + \alpha_2 = \frac{K_2}{d_2}. \tag{6.22}$$

By solving Eqs. (6.21) and (6.22), the values of α_1 and α_2 can be found as expressed in Eqs. (6.23) and (6.24).

$$\alpha_1 = \frac{1}{\sin^2\theta}\left(\frac{K_1}{d_1} - \cos\theta \cdot \frac{K_2}{d_2}\right) \tag{6.23}$$

$$\alpha_2 = \frac{1}{\sin^2\theta}\left(\frac{K_2}{d_2} - \cos\theta \cdot \frac{K_1}{d_1}\right) \tag{6.24}$$

Let us go back to consider Fig. 6.13 again, according to the Pythagoras theorem, α_3 is

$$\begin{aligned}\alpha_3^2 &= l_1^2 - \|\alpha_1\mathbf{n}_1 + \alpha_2\mathbf{n}_2\|^2 \\ \alpha_3 &= \sqrt{l_1^2 - \|\alpha_1\mathbf{n}_1 + \alpha_2\mathbf{n}_2\|^2}.\end{aligned} \tag{6.25}$$

Now we have known all values of α_i $(i = 1, 2, 3)$. Since $A_4 - A_1 = \alpha_1\mathbf{n}_1 + \alpha_2\mathbf{n}_2$, therefore the equation for computing the position of point P as shown previously in Eq. (6.11) can be rewritten as

$$P = A_1 + (A_4 - A_1) + \alpha_3\mathbf{n}_3, \quad \alpha_3 \geq 0. \tag{6.26}$$

SPIDAR-8 senses eight fingertip positions on the user's left and right hands simultaneously in real-time using Eq. (6.26) by substituting the corresponding length of strings and string fulcrums of each finger. In every update of position sensing if any fingertip position comes into contact with the virtual object, force feedback will be generated. SPIDAR-8 generates force feedback at the fingertips of the user by controlling the tension of the strings. The detail of force feedback generation is explained in the following subsections.

6.3.2 *Force Feedback Generation by 3 Strings*

SPIDAR-8 uses the resultant force composed by the tension of three strings to display force feedback at the fingertips of the user. The tension on each string is generated by controlling the amount of electric current entering each motor. Considering the diagram shown in Fig. 6.14, the resultant force vector is **f** and the unit vectors of the tension of strings are $\mathbf{u}_i$ $(i = 1, 2, 3)$. We express the resultant force as

$$\mathbf{f} = \sum_{i=1}^{3} \tau_i \mathbf{u}_i, \tag{6.27}$$

where τ_i $(i = 1, 2, 3)$ is the value of the tensions of strings.

6.3.2.1 Force Direction

Because the arrangement of three actuators on the frame and the way that three strings are connected to each fingertip of the user, the tension of string cannot be generated in every cases as shown by examples in Fig. 6.15. In a case as shown in Fig. 6.15(a), tension forces can be produced on all strings, therefore the resultant force can be composed and force can be displayed. Conversely, in the case of Fig. 6.15(b), tension cannot be produced on any string to display a force as represented by the dash arrow. This is a limitation of SPIDAR-8 in which force feedback on each finger cannot be displayed in every directions.

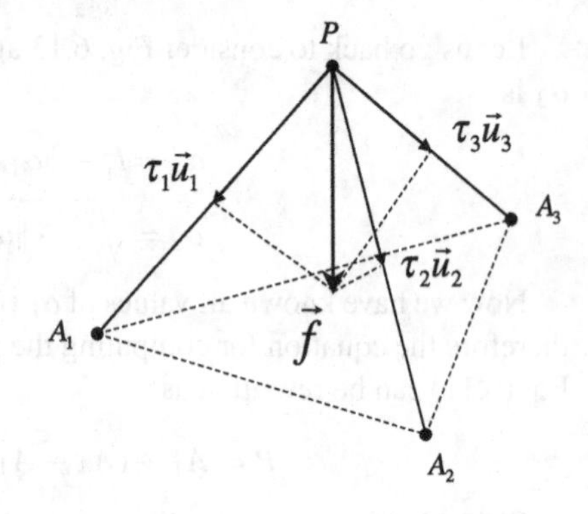

Fig. 6.14 Force feedback generation using 3 strings

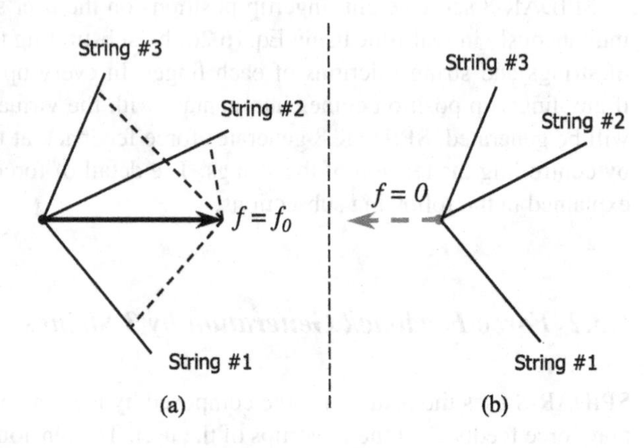

Fig. 6.15 Force display

6.3.2.2 Force Feedback Generation from Projected Force

Anther problem of displaying force of SPIDAR-8 is to display force feedback when the direction of force is outside the force cone of the finger. Force cone for each finger can be defined by three strings from three different actuators connected at a point on a fingertip. If force vector, which is computed from the collision of fingertip position and the virtual object is completely inside the force cone as in Fig. 6.15(a), SPIDAR-8 can display force completely at the user's fingertip. But if the force vector is outside the force cone, SPIDAR-8 needs to project force vector back to the force cone and re-composes the resultant force by using the tension of 1, 2, or 3 strings that force vector can be positively projected. There are three possible cases of force projection as shown in Fig. 6.16 that the resultant force can be re-composed; (a) from the projection on one string, (b) from the projection on two strings, and (c) from the projection on three strings.

The procedures of force projection can be described as follows.

Notation

- **f**: Force vector
- ϕ_i: Unit force vector along the ith string ($i = 1, 2, 3$)

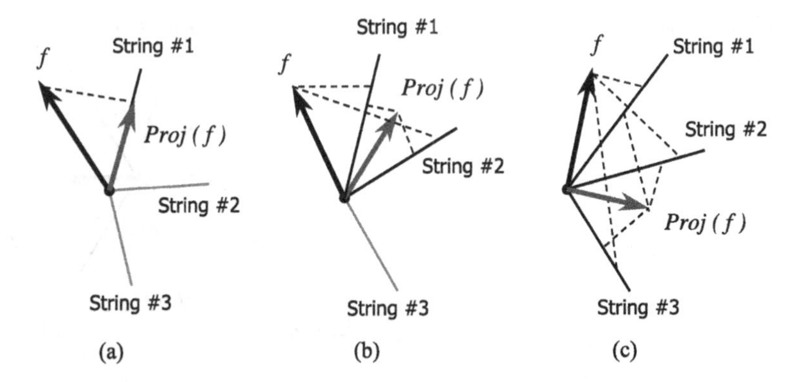

Fig. 6.16 Different cases of force projection

- τ_i: Tension force on the ith string ($i = 1, 2, 3$)
- ω_i: Projection value of force vector on the ith string ($i = 1, 2, 3$)

Step I Compute the Projection of Force Computes the projection force vector onto all strings, $\omega_i = (\phi_i, \mathbf{f})$ where ($i = 1, 2, 3$). Next, arranges ω_i in ascending order in such a way that $\omega_{i1} \geq \omega_{i2} \geq \omega_{i3}$ and then considers the value of τ in the following cases.

- $\omega_{i1} \leq 0 \Rightarrow \tau_{i1} = \tau_{i2} = \tau_{i3} = 0$. This means that no positive projection of resultant force vector on any string. Thus the tension on all strings are assigned to zero and then leaves the process.
- $\omega_{i1} \geq 0 \geq \omega_{i2} \Rightarrow \tau_{i1} = \omega_{i1}, \tau_{i2} = \tau_{i3} = 0$. Force vector, $\mathbf{f}$, can be positively projected only on ϕ_{i1}. Therefore assigns the tension of the $i\,1^{st}$ string to the projection value and the rest of the strings to zero and leaves the process.
- $\omega_{i2} \geq 0 \geq \omega_{i3} \Rightarrow \tau_{i3} = 0$ go to *Use_2_Strings*. Since both the i1st and the i2nd strings will be used, assigns the tension on the i3rd to zero and goes further to the step of Use_2_Strings.
- $\omega_{i3} \geq 0 \Rightarrow$ go to *Use_3_Strings*. In this case, all strings can be positively projected therefore the process continues in the step of Use_3_Strings.

Step II Use_2_Strings Arranges a matrix A of the unit vectors of two projected strings. Then computes the values of tension, τ', from matrix A and force vector as follow.

$$A = (\phi_{i1}, \phi_{i2}), \quad \text{where } A \in R_{2\times 3}$$

$$\tau' = \left(A^T A\right)^{-1} A^T \cdot \mathbf{f}$$

$$= \begin{bmatrix} 1 & \alpha \\ \alpha & 1 \end{bmatrix}^{-1} \begin{bmatrix} \omega_{i1} \\ \omega_{i2} \end{bmatrix}, \quad \text{where } \alpha = (\phi_{i1}, \phi_{i2}).$$

Arranges τ' in such a way that $\tau'_{i1} \geq \tau'_{i2}$ and considers the τ' in the following cases.

- $\tau'_{i1} > 0 \geq \tau'_{i2} \Rightarrow \tau'_{i1} = \omega_{i1}, \tau'_{i2} = 0$. In this case, only the i1st is used, assigns the tension of this string to the projected value and the other one to zero.

Fig. 6.17 Force space

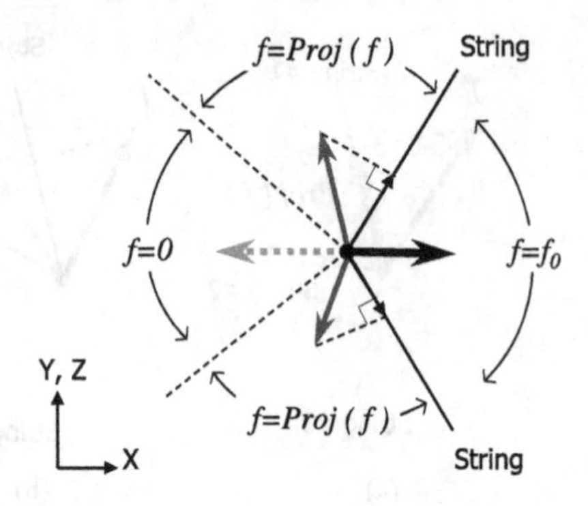

- $\tau'_{i1} > \tau'_{i2} > 0 \Rightarrow \tau_j = \min(\tau'_j, \omega_j, \tau_{\max})$. Both strings are to be used and the tension on each string is selected from the minimum value among these three values; the tension calculated within this step, the projected value from Step 1, and the predefined maximum tension value (for safety concerned).

Step III Use_3_Strings Arranges a matrix A, which elements are the unit vectors on three strings. Then computes the values of tension from matrix A and force vector from the following equations.

$$A = (\phi_{i1}, \phi_{i2}, \phi_{i3}), \quad \text{where } A \in R_{3\times 3}$$

$$\tau' = \left(A^T A\right)^{-1} A^T \cdot \mathbf{f}$$

$$= \begin{bmatrix} 1 & \alpha & \beta \\ \alpha & 1 & \gamma \\ \beta & \gamma & 1 \end{bmatrix}^{-1} \begin{bmatrix} \omega_{i1} \\ \omega_{i2} \\ \omega_{i3} \end{bmatrix},$$

$$\text{where } \alpha = (\phi_{i1}, \phi_{i2}),\ \beta = (\phi_{i1}, \phi_{i3}),\ \gamma = (\phi_{i2}, \phi_{i3}).$$

Arranges τ' in ascending order in such a way that $\tau'_{i1} \geq \tau'_{i2} \geq \tau'_{i3}$, then considers the τ' in the following cases.

- $\tau'_{i1} > 0 \geq \tau'_{i2} \Rightarrow \tau'_{i1} = \omega_{i1}, \tau'_{i2} = \tau'_{i3} = 0$. In this case, only the i1st is used, assigns the tension for this string to the projected value and the rest of the strings to zero.
- $\tau'_{i2} > 0 \geq \tau'_{i3} \Rightarrow \tau'_{i3} = 0$ go to *Use_2_Strings*. Two strings are to be used. First, assigns the tension on the ird to zero and then sends the process back to the step of Use_ 2_Strings.
- $\tau'_{i3} > 0 \Rightarrow \tau_j = \min(\tau'_j, \omega_j, \tau_{\max}), j = (i1, i2)$. The tension forces on all three strings are used. The value of the tension of each string is selected from the minimum value among these three values; the tension calculated within this step, the projected value from Step 1, and the predefined maximum tension value.

The result when applying above algorithm, SPIDAR-8 displays appropriate force in the cases of projection force vector. However, the magnitude of force feedback is reduced from the actual value because the projected tension values are used instead. We have shown a model of force space, which illustrates the areas that SPIDAR-8 can display force for a finger by Fig. 6.17.

6.4 Applications

6.4.1 Introduction

This section presents three applications by employing SPIDAR-8 as a haptic interface device. The first application is the virtual hands pose estimation from the fingertip positions measured by SPIDAR-8 in real-time [14]. The virtual hands representing the user's hands are displayed in the computer simulated virtual world. The result from the estimation show the virtual hands move according to the movement of the user's hands. In the second application, the user is allowed to manipulate the virtual objects with force feedback sensation. The manipulation of a virtual Rubik's cube is implemented. Not only force sensation that can be perceived, but the user can also see his or her manipulation of the virtual Rubik's cube by the virtual hands on the display screen. Finally, the third application is the same manipulation application of the virtual Rubik's cube. However, the significant difference is instead of displaying the virtual hands, the live image sequences of the user's real hands manipulating the virtual cube are merged into the virtual world. In this application, the method of constructing a mixed reality environment for virtual objects manipulation task with force feedback is presented. Each application briefly explained above is described in detail as follows.

6.4.2 Virtual Hands Pose Estimation

6.4.2.1 Model of Real Hand

On one of our hand, it consists of many parts and joints. The word *parts* is referred to the palm and the finger links. While the word *joints* is referred to finger joints found either between two finger links or between finger link and the palm. We can change different postures of our hand by controlling the movements at finger joints. The movements at each finger joint is defined by the value of degree of freedom (DOF) at that joint.

Figure 6.18 shows a simple structure of a hand labeled with number of DOF at each joint and at the wrist. In total, there are twenty-six DOFs found on a hand in which twenty of them are at all finger joints and six DOFs are the translation and rotation at the wrist.

6.4.2.2 Model of Virtual Hand

In this application, SPIDAR-8 is used and only eight fingertip positions on both hands of the user are measured. Therefore on each hand, there are only four three-dimensional position information that is equivalent to twelve elements of information. With the available information, it is difficult to model a hand of twenty-six

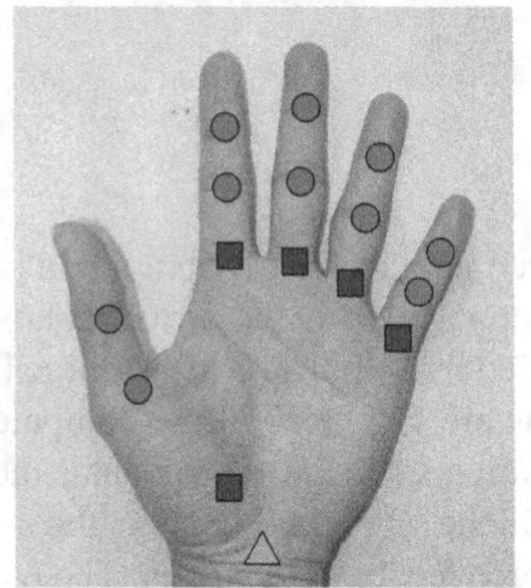

Symbol	Degree of freedom
●	1
■	2
△	6
Total	26

Fig. 6.18 Model of real hand

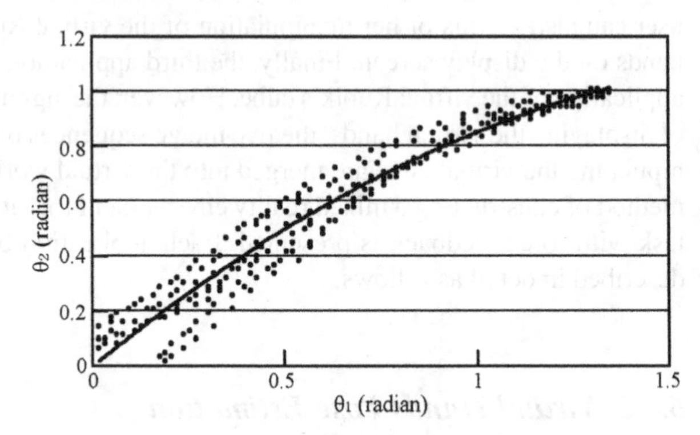

Fig. 6.19 Relationship of joint angle closed to the fingertip and the next joint on the same finger

DOFs. To make it possible, the DOF on the model of virtual hand should be reduced. There are three criterions for reducing DOF, which are associated to the way that position information are provided by SPIDAR-8.

1. Since the little finger is not measured by SPIDAR-8, therefore the motion of this finger is not taken in the computation. However, its motion follows the motion of the ring finger in each corresponding joint.
2. In most cases, we are almost impossible to move the joint closed to the fingertip without moving the next joint without forcing the finger to move in unnatural way. This is because there is a dependency between these two joints in which causes by the same tendon used for moving inside the finger. Therefore, the DOF at the joint closest to the fingertip can be omitted in the computation. Instead, its motion is given in proportional to the motion of next joint.
3. Since the middle finger is not move side by side in most cases when moving the fingers, thus the DOF at the joint between the finger link of the middle finger and the palm can be reduced from two to one DOF.

The relationship of two joint angles can be found by measuring the bending angle of both joint angles several times. Figure 6.19 shows a graph of bending angles of both joints θ_1 and θ_2 (refer to Fig. 6.20). From the graph, a second-degrees polyno-

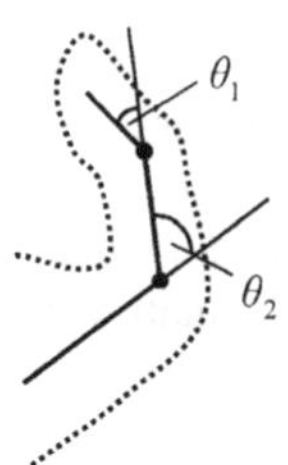

Fig. 6.20 Two joint angles on a finger

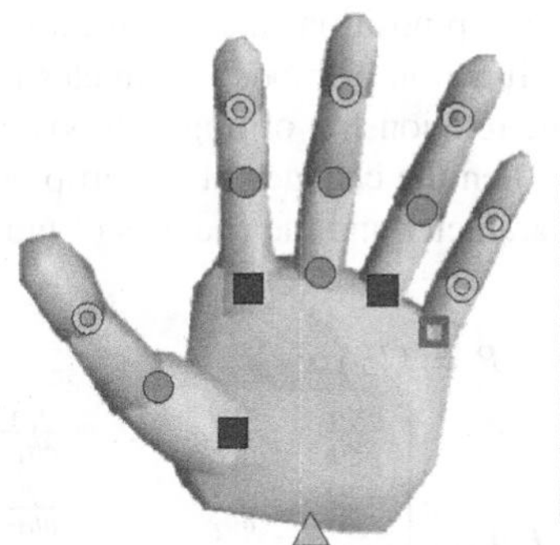

Symbol	Degree-Of-Freedom
●	1
■	2
△	6
◎	●
□	■
Total	17

Fig. 6.21 Model of virtual hand; reduced DOFs

mial equation as in Eq. (6.28) can be approximated. This relation can make the joint closed to the fingertip move in proportional to the next joint on the same finger.

$$\theta_2 = 1.1341\theta_1^2 - 0.286\theta_1 \tag{6.28}$$

In summary, four DOFs can be reduced after the first criterion, another four DOFs can be reduced after the second criterion, and one DOF can be reduced after the third criterion. As shown in Fig. 6.21, the model of a virtual hand has in total seventeen DOFs in which it can be estimated by position information provided by SPIDAR-8.

6.4.2.3 Hands Pose Estimation Algorithm

Algorithm for estimating a virtual hand pose can be described as follows.

Step 1. Takes fingertip positions of thumb, index finger, middle finger, and ring finger on each hand measured by SPIDAR-8.
Step 2. Computes the changes between fingertip positions measured by SPIDAR-8 and the fingertip positions on the virtual hand.
Step 3. If there is no change of position, the estimation needs not to be performed. So the process can leave the algorithm, otherwise the process continues to the next step.
Step 4. Computes the changes of joint angles from the changes of fingertip positions and adjust all changes, fingertip position and joint angles, to model of virtual hand.
Step 5. Repeats the processes again from Step 2.

The computation of changes of joint angles in Step 4 of the algorithm can be elaborated as follows.

Let express fingertip positions as matrix P and model parameters of a virtual hand, which consists of all joint angles and the position and orientation of the wrist, be matrix θ as shown in Eqs. (6.29) and (6.30) respectively.

$$P = (p_1, p_2, \ldots, p_{12})^t \tag{6.29}$$

$$\theta = (\theta_1, \theta_2, \ldots, \theta_{17})^t \tag{6.30}$$

It can be said that fingertip positions are the result of function of model parameters as in Eq. (6.31). If function of model parameter is substituted by a Jacobian matrix $J(\theta)$, which is the relationship of fingertip positions and model parameters as shown in Eq. (6.32), Then the changes of fingertip positions can be found from the Jacobian of model parameter and the changes of model parameter as shown by Eq. (6.33).

$$P = f(\theta) \tag{6.31}$$

$$J(\theta) = \begin{bmatrix} \frac{\partial p_1}{\partial \theta_1} & \frac{\partial p_1}{\partial \theta_2} & \cdots & \frac{\partial p_1}{\partial \theta_{17}} \\ \frac{\partial p_2}{\partial \theta_1} & \frac{\partial p_2}{\partial \theta_2} & \cdots & \frac{\partial p_2}{\partial \theta_{17}} \\ \vdots & \vdots & \ddots & \vdots \\ \frac{\partial p_{12}}{\partial \theta_1} & \frac{\partial p_{12}}{\partial \theta_2} & \cdots & \frac{\partial p_{12}}{\partial \theta_{17}} \end{bmatrix} \tag{6.32}$$

$$dP = J(\theta)d\theta \tag{6.33}$$

In another word, the changes of model parameters can be computed by the pseudo-inverse matrix of the Jacobian and the changes of fingertip positions as in Eq. (6.34). The virtual hand pose can be estimated by adjusting the changes of fingertip positions and joint angles to the model of virtual hand.

$$d\theta = J(\theta)^+ dP \tag{6.34}$$

$$J(\theta)^+ = \left(J(\theta)^t J(\theta)\right)^{-1} J(\theta)^t \tag{6.35}$$

6.4.2.4 Results of the Estimation

Figure 6.22 shows a photograph of virtual hands pose estimation by SPIDAR-8. The virtual hands displaying on the screen can move according to the movement of the user's hands. The results of virtual hand pose estimation are also shown in Fig. 6.23. In this figure, each virtual hand pose is estimated according to the corresponding posture of real hand.

6.4.3 Virtual Object Manipulation

SPIDAR-8 is employed as the interface device for the manipulation of the virtual objects in the virtual world. Force feedback can be realize during the manipulation.

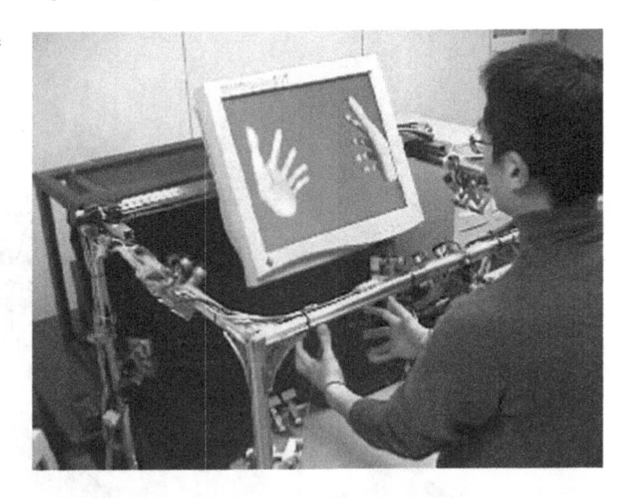

Fig. 6.22 Virtual hands pose estimation by SPIDAR-8

As visual feedback, the virtual hands, which move according to the movement of the user's hands, are displayed in the scene of virtual world. By providing with both force and visual feedbacks, the user can manipulate the virtual objects effectively.

6.4.3.1 System Setup

The arrangement of hardware equipment for the virtual object manipulation is shown by a diagram in Fig. 6.24.

An 18-inches LCD flat-screen display is installed on top of the frame of SPIDAR-8. The LCD display is set incline by 60 degree for comfortable view with less reflection of room-light. The frame is placed on a table about 70 cm high. The user is standing in front of the frame when performing the manipulation. The fingertips on both hands of the user are worn by finger caps attached by strings. The space enclosed by the frame is the space where the user can manipulate the virtual objects. The user can see the simulated virtual world, which consists of the virtual objects and the virtual hands, on the LCD display.

The user stands in front of SPIDAR-8 when performing the manipulation task. The change of system setup from sit-down configuration to stand-up configuration has improved the usability of the system in the following issues.

- *Reducing user fatigue.* By sitting in front of the frame and perform the task, the user needs to raise up and forward his or her arms into the middle the frame. For the manipulation space that the hands have to be raised up higher than the elbows, it quickly makes the user get muscular fatigue on both shoulders and upper arms. On the other hand, the user needs not to raise the hands in the stand-up posture, thus the muscular fatigue on the shoulders and upper arms are less. The user can perform the manipulation in longer period of time.
- *Avoiding the occlusion of real hands with the view of the virtual world.* We place a large-screen monitor behind the frame in the sit-down configuration for displaying the scene of the virtual world. In this case, the user see not only the virtual

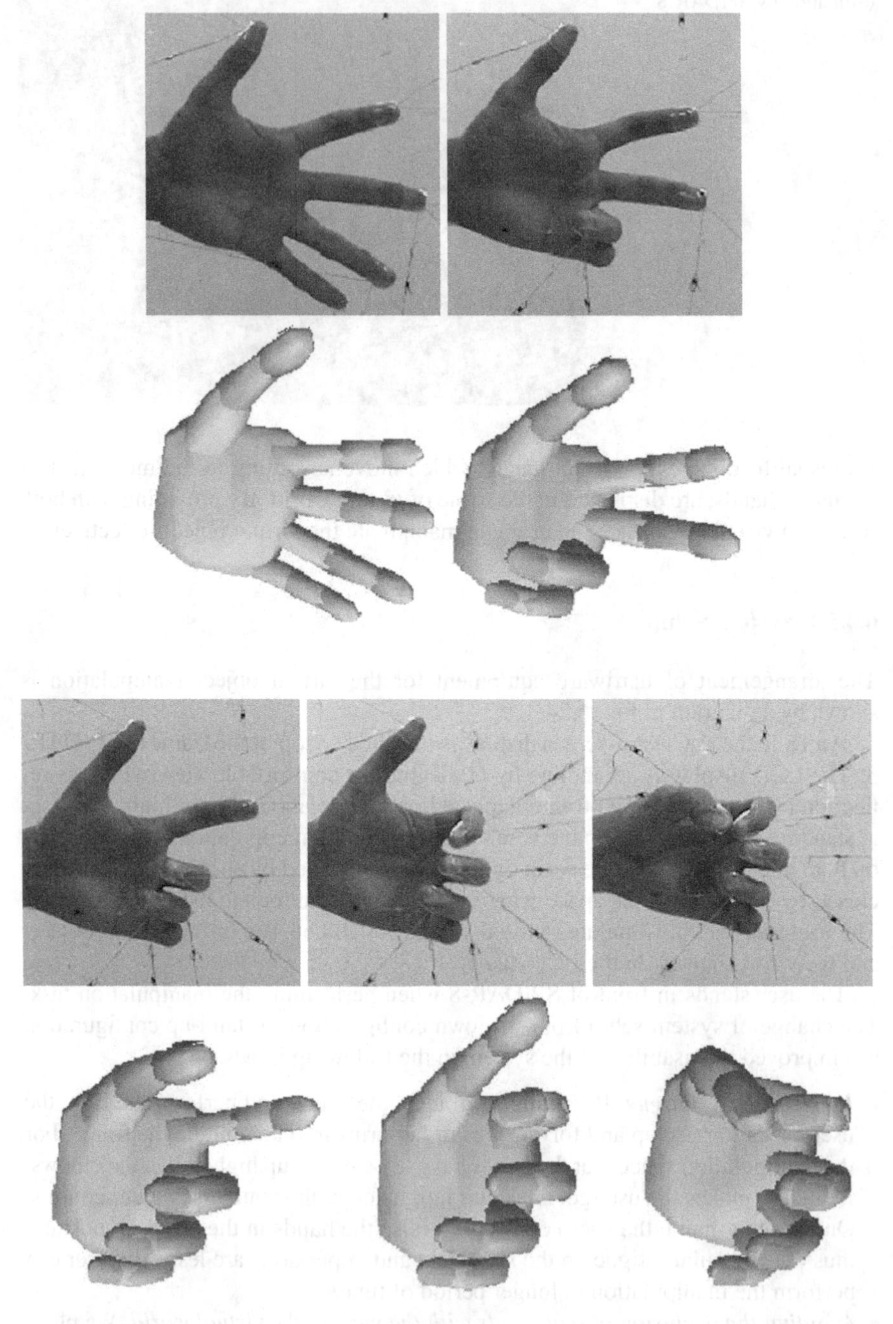

Fig. 6.23 Results of virtual hand pose estimation

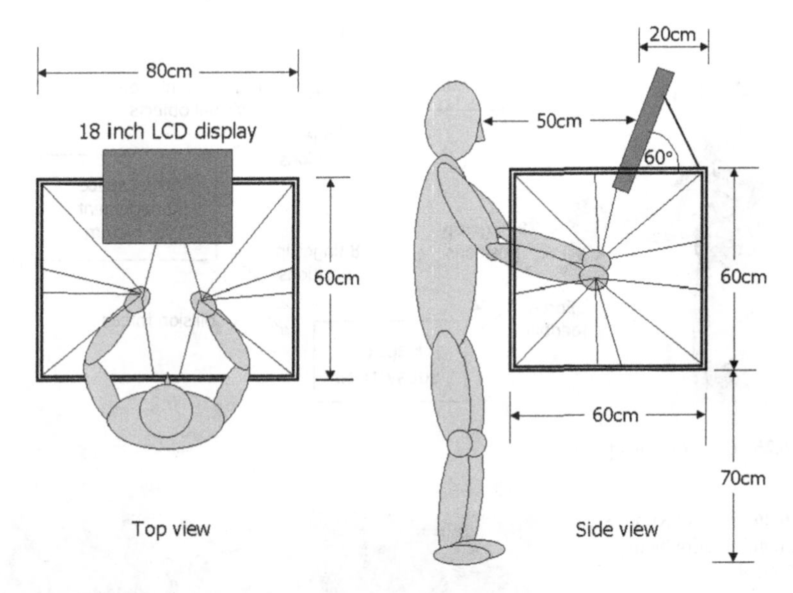

Fig. 6.24 System overview

hands but also his or her real hands. In many cases, the real hands of the user often occlude the view of manipulation. The large screen monitor at the back is removed and installs a LCD display installed on top of the frame, the problem of occlusion is solved.

6.4.3.2 System Control

A personal computer, the haptic server, is already used for controlling SPIDAR-8. Since the process of virtual hands pose estimation requires computational power, one more PC is added. The functions of the second PC are estimation virtual hands pose and rendering the virtual world. This PC is referred as the graphics server. Both PCs are connected together by computer network. Figure 6.25 shows the control concept of the system.

The whole system is divided into three subsystems as

1. Haptic subsystem
2. Virtual world management subsystem
3. Visual subsystem

In the haptic subsystem, SPIDAR-8 measures eight fingertip positions on the user's real hands. The measured fingertip positions are sent to the virtual world management subsystem. In the virtual world management subsystem, the collision of fingertip positions and the virtual objects is detected and the collision force of each finger is computed and sends back to haptic subsystem for displaying. The motion of the virtual objects are depended on the force exerted by each finger. In

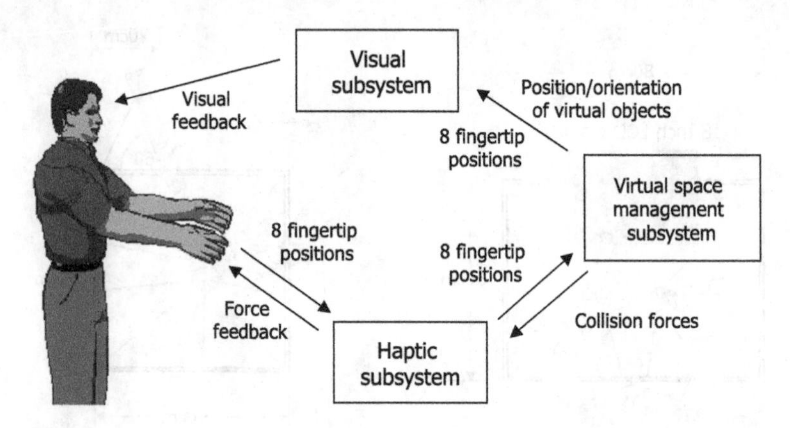

Fig. 6.25 Control concept

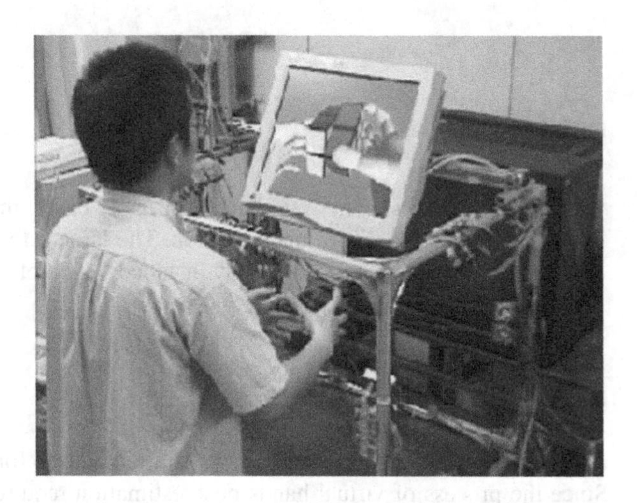

Fig. 6.26 Virtual Rubik's cube manipulation using SPIDAR-8

the visual subsystem, the virtual hands and the virtual objects are updated. Then the virtual world is rendered to the display where the user can see the performing task.

6.4.3.3 Virtual Rubik's Cube Manipulation

Figure 6.26 shows a picture of a user is manipulating of a virtual Rubik's cube using SPIDAR-8. The virtual Rubik's cube is simpler than a real one. It has $2 \times 2 \times 2$ cells in which the column cells or the row cells can be rotated in the same way as the real Rubik's cube.

This application can clearly confirm the abilities of SPIDAR-8. To grasp and rotate cells of the cube, dexterous manipulation of multi fingers and the use of both hands are required. It needs the fingers on one hand to grasp on one side

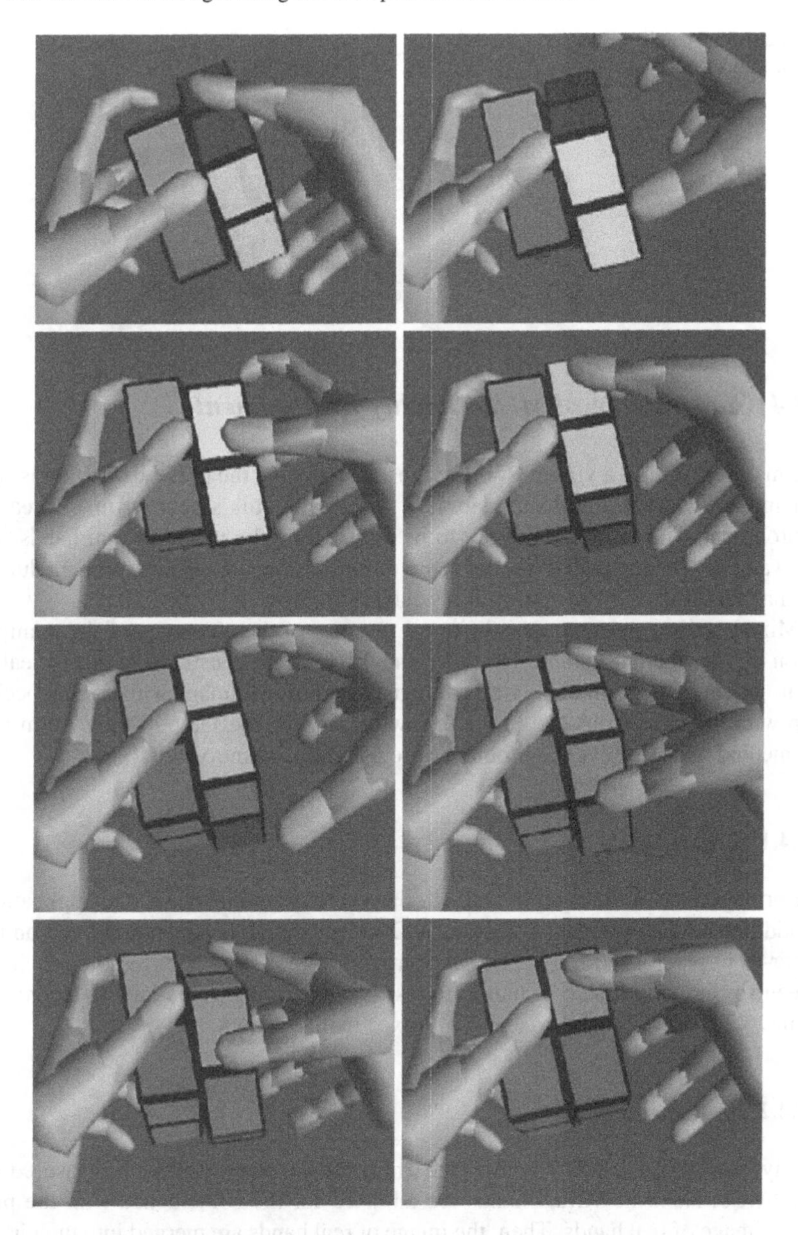

Fig. 6.27 Screens of virtual Rubik's cube manipulation

of the cube and the fingers on another hand to grasp on the opposite side, then the column cells or the row cells can be rotated. However, by only single hand grasping the cube, the whole cube is manipulated. Figure 6.27 shows some captured images of the manipulation of virtual Rubik's cube shown on the display screen.

Fig. 6.28 Real hands versus virtual Rubik's cube

6.4.4 Construction of Mixed Reality Environment

The manipulation of a virtual Rubik's cube described in the previous section is further implemented to the manipulation of a virtual Rubik's cube in mixed reality environment. The virtual hands are to be replaced by the image of real hands. By this way, the user can realized better sense of immersion into the environment during the manipulation of the virtual Rubik's cube (Fig. 6.28).

Mixed reality environment is believed to improve the efficiency of the manipulation in virtual environment. However, the difficulty of constructing mixed reality environment is problem of solving the registration of real image with correct occlusion with the virtual objects. In the following section, the setup of the system and the method used to solve the problem of registration are introduced.

6.4.4.1 System Setup

The arrangement of the equipment is similar to the system in the second application. In addition to the previous system, a CCD video camera is added and fixed on the upper-front pipe of the frame. The background of the frame is covered by the blue screen so that only image of hands are extracted by the chroma-keying technique. Figure 6.29 shows control concepts of mixed reality system.

6.4.4.2 Implementation

Firstly, the virtual hands are modeled from the fingertip positions measured by SPIDAR-8. Next the virtual hands are adjusted for the coincidence with the projected image of real hands. Then, the image of real hands are merged into the virtual world by superimposing on the transparent virtual hands. By this way, the correct occlusion of real hands and the virtual objects can be achieved. However, the process of adjusting model of virtual hands requires position information of the hand in detail. Only four fingertip positions on each of the user's hand are measured, the information are insufficient to solve the registration problem. Therefore, additional position information of the hand are extracted. Two items are established to provide additional position information, i.e., the projected wrist position and the joint ball.

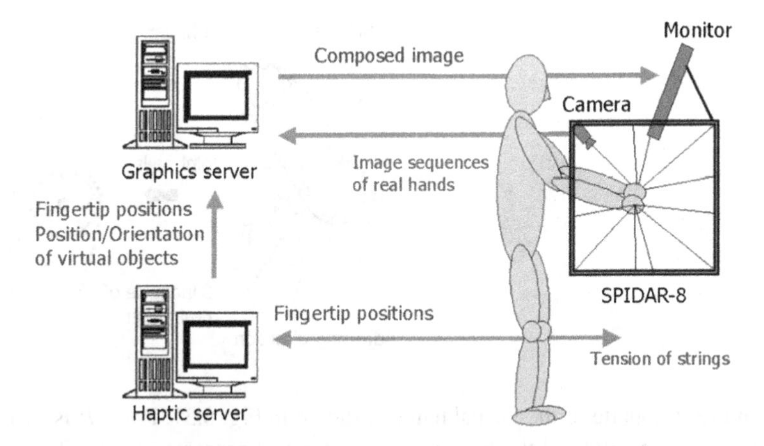

Fig. 6.29 Control concept of mixed reality system

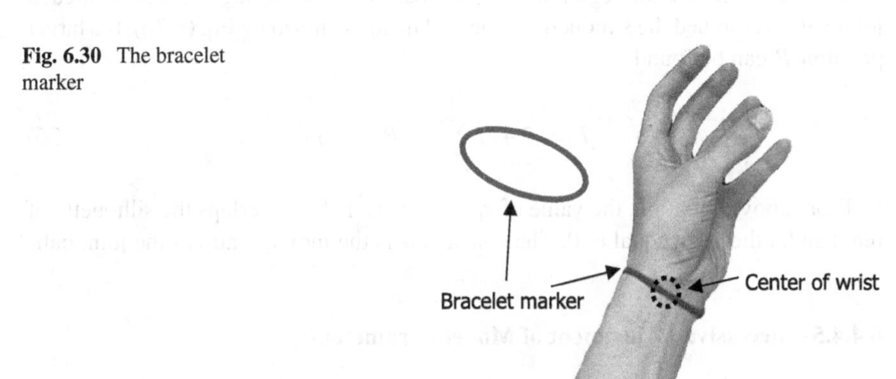

Fig. 6.30 The bracelet marker

6.4.4.3 Projected Wrist Position

A projected wrist position can be found from screen data when image of real hand taken by the camera is displayed on the screen. User is asked to wear a red bracelet marker on the wrist as shown in Fig. 6.30. The screen data is raster-scanned and then the marker can be extracted by the color-filtering technique. From both ends of the extracted marker, a position at the center is computed. This position represents a projected wrist position on one hand.

6.4.4.4 Joint Ball

In many cases the projected images of real hands are not completely coincide with the virtual hands, especially at the finger joints. Such errors can be corrected by adjusting the finger joints of the virtual hands. By considering the silhouettes of real hand and the virtual hand on the screen data, a joint ball is defined at each finger

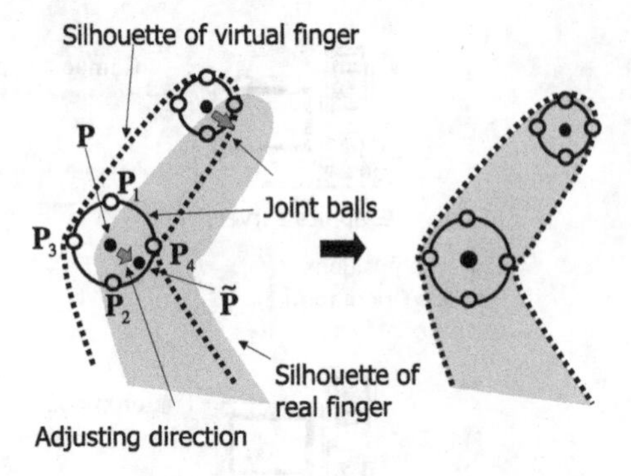

Fig. 6.31 Joint ball

joint on the silhouette of the virtual hand as shown in Fig. 6.31. Point P is assigned to be the initial position of the finger joint and four adjacent points P_1, P_2, P_3, and P_4 are used to define the region of the joint ball. By considering the region of each joint ball overlapped the silhouette of the real hand as shown by Eq. (6.36), the target position $\tilde{P}$ can be found.

$$\tilde{P} = P + \alpha \sum_{i=1}^{4} \tau_i (P_i - P) \tag{6.36}$$

From above equation, the value of τ_i is equal to 1 if P_i overlaps the silhouette of real hand, otherwise equal to 0. The constant α is the moving ratio of the joint ball.

6.4.4.5 Successive Adjustment of Model Parameters

In this application, a model of real hand consists of four fingertip positions measured by SPIDAR-8 and the projected wrist position. While a model of virtual hand is constructed with the initial fingertip positions and wrist position. If all parameters of both models are equal, the projected image of real hand can be completely coincide with the virtual hand. However, both of them are always different. The model of virtual hand is adjusted in every update. The adjusting amounts can be found from model of real hand at current step and the approximately adjusted model of virtual hand in the next step. The successive adjustment can be described as follows.

Let model parameters of virtual hand being defined as

$$Q = (Q_1, Q_2, Q_3, Q_4, Q_5) \tag{6.37}$$

where Q_1, Q_2, Q_3, and Q_4 are fingertip position of thumb, index, middle, and ring finger respectively and Q_5 is a position at the wrist.

In the same way, model parameters of real hand being defined as

$$\tilde{Q} = (\tilde{Q}_1, \tilde{Q}_2, \tilde{Q}_3, \tilde{Q}_4, \tilde{Q}_5) \tag{6.38}$$

where $\tilde{Q}_i$ $(i = 1, 2, 3, 4)$ are measured fingertip positions by SPIDAR-8 and $\tilde{Q}_5$ is a projected wrist position.

The initial positions of joint balls are

$$P = ((x_1, y_1), (x_2, y_2), \dots, (x_k, y_k)), \tag{6.39}$$

and the target positions of joint balls are

$$\tilde{P} = ((\tilde{x}_1, \tilde{y}_1), (\tilde{x}_2, \tilde{y}_2), \dots, (\tilde{x}_k, \tilde{y}_k)) \tag{6.40}$$

The differences of model parameters, ΔQ, can be computed from model parameters of real hand and model parameters of virtual hand as

$$\Delta Q = \tilde{Q} - Q \tag{6.41}$$

The differences of position of joint balls, ΔP, can be found from the initial and the target position of the joint balls as

$$\Delta P = \tilde{P} - P \tag{6.42}$$

The integrated error, J, is defined by the differences of model parameters, ΔQ, and the differences of finger joints, ΔP with a relative weighting coefficient c between ΔQ and ΔP.

$$J = \|\Delta Q\|^2 + c\|\Delta P\|^2 \tag{6.43}$$

If the integrated error is small enough, model parameters of real hand and model parameters of virtual hand are assumed to be the same. The result is the projected image of real hand is coincident with the virtual hand.

In the following algorithm, the adjustment of model parameters θ at the nth step to be model parameters in the next step, the nth $+ 1$ step, the adjusting amounts can be computed based on optimization criterion as follow.

Step 1. Constructs model parameters of real hand, $\tilde{Q}$, as in Eq. (6.38) from the updated fingertip positions measured by SPIDAR-8 and the extracted projected wrist position.
Step 2. Computes the differences of positions of joint balls by Eq. (6.42).
Step 3. Computes the differences of model parameters ΔQ by Eq. (6.41) and the differences of joint balls ΔP by Eq. (6.42).
Step 4. Considers the integrated error, J, if the integrated error is small enough, two model parameters are assumed to match with each other, so the process finishes and exits the algorithm. Otherwise, the process continues in the next step.
Step 5. Computes the adjusting amounts, $\Delta\theta$, and adjusts model parameters θ_n to be model parameters θ_{n+1}.
Step 6. Repeats the algorithm from Step 2.

In the following, the computation of the adjusting amounts $\Delta\theta$ in Step 5 is described.

Generally, the adjusting amounts, $\Delta\theta$, is the differences between model parameters in the next step, nth $+ 1$, and model parameters at the current step, nth.

$$\Delta\theta = \theta_{n+1} - \theta_n \tag{6.44}$$

As we have model parameters $\tilde{Q}$, model parameters of the nth + 1 step should be $Q(\theta_{n+1})$. Then, the integrated error can be expressed as

$$J(\theta_{n+1}) = \left\| Q(\theta_{n+1}) - \tilde{Q} \right\|^2 + c \left\| P(\theta_{n+1}) - \tilde{P} \right\|^2 \tag{6.45}$$

Model parameters $Q(\theta_{n+1})$ can be linearly approximated as the model parameters $Q(\theta_n)$ with the adjusting amounts that can be found from the Jacobian matrix $\mathbf{J}_Q$.

$$\begin{aligned} Q(\theta_{n+1}) &= Q(\Delta\theta + \theta_n) \\ &\cong Q(\theta) + \mathbf{J}_Q(\theta_n)\Delta\theta \end{aligned} \tag{6.46}$$

The Jacobian metric $\mathbf{J}_Q(\theta_n)$ is

$$\mathbf{J}_Q(\theta_n) = \frac{\partial Q(\theta_n)}{\partial \theta_n} \tag{6.47}$$

In the same way, we can linearly approximate the target positions of joint balls $P(\theta_{n+1})$ as

$$P(\theta_{n+1}) \cong P(\theta_n) + \mathbf{J}_P(\theta_n)\Delta\theta \tag{6.48}$$

Therefore we can linearly approximate the integrated error in Eq. (6.45) by the approximated of $Q(\theta_{n+1})$ and $P(\theta_{n+1})$ from Eqs. (6.46) and (6.48) respectively as

$$\begin{aligned} J(\theta_{n+1}) \cong& \left\| \mathbf{J}_Q(\theta_n)\Delta\theta - \left(\tilde{Q} - Q(\theta_n)\right) \right\|^2 \\ &+ c \left\| \mathbf{J}_P(\theta_n)\Delta\theta - \left(\tilde{P} - P(\theta_n)\right) \right\|^2 \end{aligned} \tag{6.49}$$

We can solve for $\Delta\theta$ by rearranging above second-degrees polynomial equation in a simple form of

$$\mathbf{A}\,\Delta\theta = \mathbf{b} \tag{6.50}$$

$$\Delta\theta = \mathbf{A}^{+}\,\mathbf{b} \tag{6.51}$$

where

$$\mathbf{A} = \mathbf{J}_Q(\theta_n)^T \mathbf{J}_Q(\theta_n) + c\mathbf{J}_P(\theta_n)^T \mathbf{J}_P(\theta_n) \tag{6.52}$$

$$\mathbf{b} = \mathbf{J}_Q(\theta_n)^T\left(\tilde{Q} - Q(\theta_n)\right) + c\mathbf{J}_P(\theta_n)^T\left(\tilde{P} - P(\theta_n)\right) \tag{6.53}$$

6.4.4.6 Merging Image of Real Hands into Virtual World

A CCD video camera is used to record image sequences of user's real hands when performing the manipulation task. The chroma-keying technique is used to extract only image of the hands from the blue screen background. We implement a software routine for the chroma-keying on the graphics server. After model of virtual hands had been adjusted as already described in the previous sections, the color pixels of the virtual hands are blitted to be transparent on the screen. From three-dimensional position information of model of virtual hands, the virtual hands always occlude the

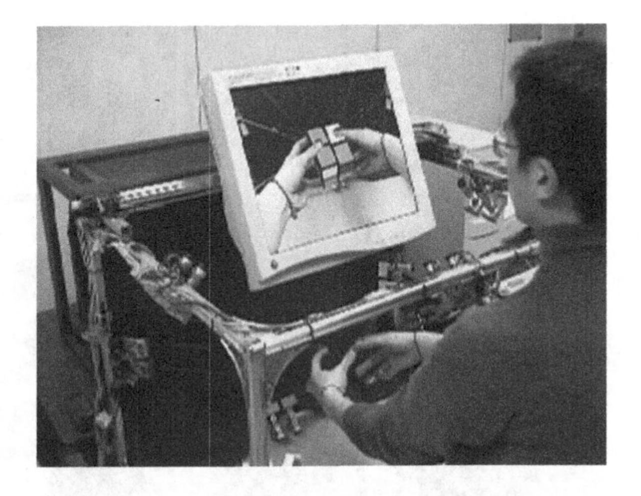

Fig. 6.32 SPIDAR-8 provides mixed reality environment

virtual objects with correct geometrical position. Then the extracted image of real hands by chroma-keying are superimposed on the transparent virtual hands. In the scene of the virtual world, image of real hands can be merged with correct occlude with the virtual objects.

Figure 6.32 shows the constructed mixed reality environment where the user is manipulating the virtual Rubik's cube with force sensation and see his real hands in the virtual world performing the with the virtual cube.

6.4.4.7 Results and Discussion

The difference of the manipulation of the virtual Rubik's cube in the mixed reality environment is the user can image of his or her real hands manipulate the cube instead of the virtual hands. In Fig. 6.33, a series of screens of the manipulation of the virtual Rubik's cube in the mixed reality environment are shown.

Some errors can be noticed from the results shown in the previous section. In some cases, the space between fingers and the virtual cube could be seen, and in some cases, some part of the finger was covered by the virtual cube. Those errors are suspected to cause by two main deflects. Firstly, the insufficient performance of the virtual hands pose estimation. It is very difficult to estimate virtual hands pose to be exactly the same as the real hands. In additional to the limited measured position information from the interface device, the additional position information such as joint balls and the projected wrist position are employed. The estimation was noticed to be improved, however some error still remain. Secondly, the insufficient of computational power on the graphics server. Since there are many heavy computational processes such as the estimation of virtual hands pose, solving the registration, and rendering the virtual world, the graphics server should be replaced with a higher performance computer.

Fig. 6.33 Screens of mixed reality environment

6.5 SPIDAR-Hand System Using Augmented Reality for Evaluating Controllability of Handheld Devices

Recently, virtual reality techniques are widely used in various areas such as game, manufacture, education, medicine and so on. Especially, those are used as methods of evaluation and testing as well as designing for a new product as CAE in the field

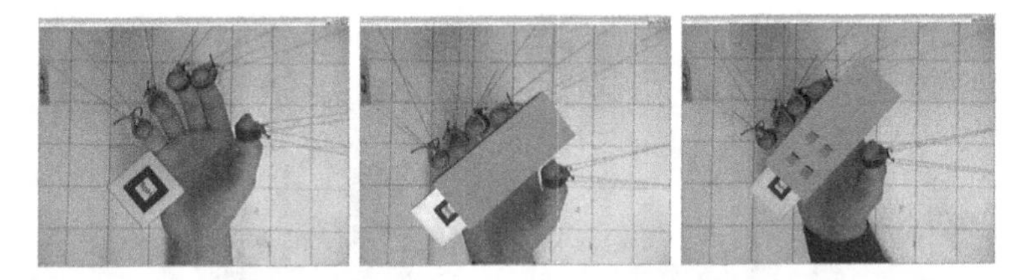

Fig. 6.34 The process of the design for the handheld device: (**a**) Detecting the marker; (**b**) Attaching the panel; (**c**) Arranging buttons

of production. For designing new devices, the producer has to consider the usability or controllability by the evaluation and needs to investigate human factors according to characteristics of the product. But, it is difficult for them to consider all of the factors affecting the possible situation, particularly those related to tactile factors. In this section, we present a new system for designing and evaluating the user interface of handheld devices, that is the SPIDAR-Hand (Space Interface Device for Artificial Reality simulating Handheld Device Controllability) system.

6.5.1 Handheld Device Design

Using our system, we can simply design a handheld device and buttons [15]. Our concept of the handheld device design for the controllability evaluation can be explained by three steps as shown in Fig. 6.34. Firstly, the SPIDAR-Hand system recognizes the marker on the mock-up and then detects the marker position. Secondly, a panel is superimposed on the marker. And lastly, designer arranges some buttons on the panel. We arranged buttons considering the relative position from the marker that is located not to be hidden by the finger. We attached the marker on the lower part of the palm in the near place of the pinkie as shown in Fig. 6.34(a).

This has two reasons as follows:

(i) For avoiding the situation of shading or covering up the marker with the fingers.
(ii) For protecting the unnatural perception of the size of the mock-up when user grasps the virtual device.

Designers can use 3D models for a panel and some buttons by loading those created previously by a design tool or selecting those programmed in advance by using OpenGL API. Figures 6.35 and 6.36 show the design method for various types of a handheld device.

It can be used for virtual devices of various shapes, sizes, and types by reforming pre-registered 3D models. In the case of a button, designer can decide properties such as the size, the shape, the number of buttons, and so on as well as the arrangement of buttons. For the arrangement of buttons, we calculated the position using the size of the button and the interval between buttons on condition that the size of each button is the same. We can automatically arrange buttons by the direction of the position X or the position Z on the world coordinate system using Eqs. (6.54) and (6.55).

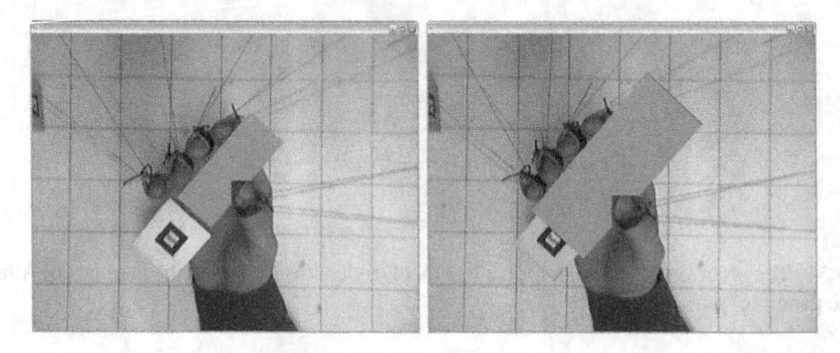

Fig. 6.35 (a) Change the size of the virtual handheld device; (b) Arrange buttons changing the number of buttons or the shape of buttons

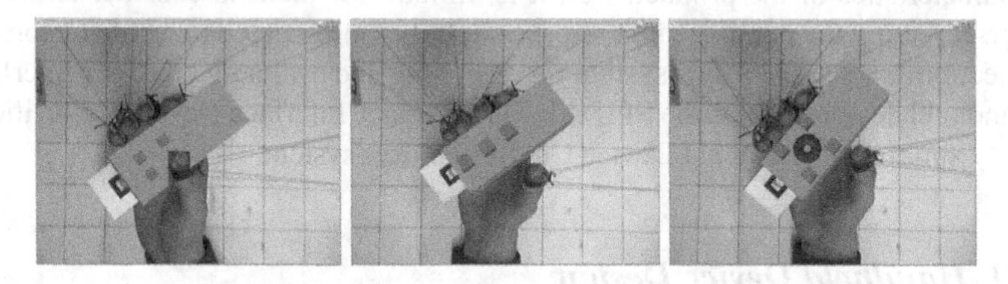

Fig. 6.36 (b) Arrange buttons changing the number of buttons or the shape of buttons; The method of the handheld device design

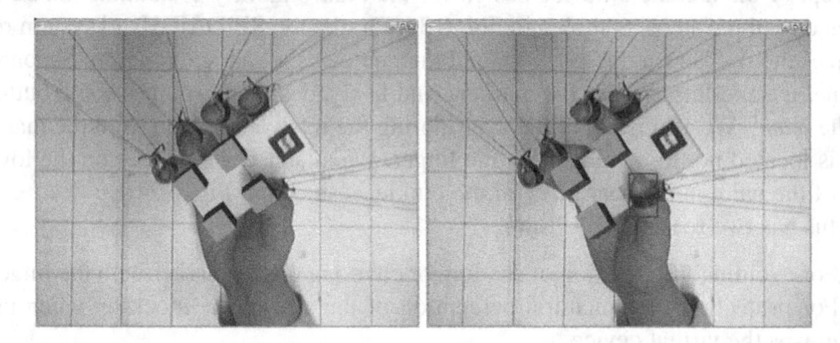

Fig. 6.37 The method of generating realistic occlusion by hiding 3D models behind objects in the real image: (a) Unrealistic occlusion; (b) Realistic occlusion

$$+direction : interval - (interfal - button\_size)/2 \tag{6.54}$$

$$-direction : -interval + (interfal - button\_size)/2 \tag{6.55}$$

After designing a virtual handheld device, developer can evaluate the new device with these models by altering their properties interactively. For giving more reality to users, we have to solve discordance on the view of overlaying the images of the camera and the graphics, particular for the phenomenon called occlusion in which an opaque object closer to the user obstructs the view of an object farther away, as shown in Fig. 6.37(a).

So, we discriminated whether it is interposition or not as follows:

```
IF
    it happens the collision
                between fingers and 3D models THEN
IF
    the upper side of the button
  < the position of the finger
THEN occlusion happens
    IF
        the lower side of the panel
      > the position of the finger
    THEN occlusion does not happen
```

Using this algorithm, the occlusion check is applied to each finger. Firstly, we check the collision between each finger and 3D models. If the collision occurs, we inspect the position of the finger and the object to know which one is closer to the user. When users grasp the handheld device, they just come to see their fingers because their palm is hidden by the device, and the button is nearer to the user than the panel. So, we compare the position of the fingers and the position of 3D models. We solved the unrealistic situation of fingers occluded by buttons introducing the blending effect using the mask image for covering the area of the fingertip when such occlusion happens as shown in Fig. 6.37(b). In addition to these graphical techniques, we need the haptic techniques to generate force feedback for giving the tactile feeling of operating the device. For testing the controllability of the devices, we have to generate two kinds of forces that are reactions against the force for grasping a handheld device and the force for pressing a button. We are discriminating these two types of force-feedback and generating them using the same way of the force transmission using the set of strings.

To make an analysis of the force feedback for clicking buttons by the tension of strings, we performed the experiment on the suitable amount of haptic feedback. With this experiment, we confirmed that our system generated the sufficient amount of force, that is 0.6 N to give the user the feeling of grasping the device, and the force from 0.5 N to 1.2 N for the feeling of clicking the button.

6.6 Conclusion

This chapter describes the proposal of two handed multi-finger haptic interface device. The device has been developed as an interface device for the interaction with the virtual reality environment.

Section 6.1 is the introductory of this chapter. We have mentioned about string-based haptic interface device SPIDAR. The SPIDAR is abbreviated from the SPace Interface Device for Artificial Reality. We introduced the SPIDAR family.

Section 6.2 describes about the system design and construction of the propose device. The design specification of hand haptic interface device has been described. We have set up the design policy based on the design specification and the user's requirement of hand haptic interface device. The proposed device is developed according to the design policy. The construction of the proposed device have been presented in detailed about hardware structure, actuator used, and the interface of the device to the user's hands.

Section 6.3 elaborates about the control of the system. The method of position sensing, which calculates 3D fingertip position of each finger of the user from the length of three strings, has been explained. Also the method of generating force feedback using the tension of strings is described. However, by using three strings, it is insufficient to compose force feedback in all of three dimensional direction. The arrangement of string attachment is also limited force display volume of each finger. Therefore, we have explained the method of projecting force used to overcome the limitation of force display of the proposed device.

In Sect. 6.4, three applications employing the proposed device had been realized. The first application was the virtual hands pose estimation. The virtual hands had been modelled and estimate their pose from the measured fingertip positions using the proposed device. The virtual hands were rendered in the virtual world in which they were moved according to the movement of the user's real hands. The user could correctly realized corresponding hand's positions in the virtual worlds. Natural and effective manipulation of the virtual objects was achieved. Secondly, the proposed device was applied to the manipulation of virtual Rubik's cube. The user was provided with force feedback in which he could manipulate the virtual Rubik's cube as in the real world. The virtual hands provided visual cues to the user during the manipulation the virtual Rubik's cube in the virtual world. The third application of the proposed device was the construction of mixed reality environment. In this mixed reality environment, not only force feedback could be realized when manipulating the virtual objects, the user could watch the live image of user's hands merged in the virtual world. The difficulty of constructing this environment was the registration of real and virtual images with correct geometrical occlusion. Although, some error of registration was incompletely solved, the method of computation introduced had shown a feasibility of applying the proposed device to the construction of the mixed reality system.

In Sect. 6.5, we are developing an AR-based 3D User Interface Design System that can measure the three dimensional position of the fingertips and give the force feedback against the fingertips while operating a virtual handheld device. The SPIDAR-Hand system that is being developed for the system, gives the force feedback to all of the five fingertips of the single hand using tensions of the strings controlled by sixteen motors, and the fingertip positions were measured by the string length and images by one camera. Our system can detect each fingertip and the hand movement in real time and generate the haptic for clicking a button and grasping a device by controlling the tension of strings.

References

1. Hirata, Y., & Sato, M. (1992). 3-dimensional interface device for virtual work space. In *Proceedings of the 1992 IEEE/RSJ international conference on IROS* (Vol. 2, pp. 889–896).
2. Ishii, M., & Sato, M. (1994). 3D spatial interface device using tensed strings. In *PRESENCE-teleoperators and virtual environments* (Vol. 3, No. 1, pp. 81–86). Cambridge: MIT Press.
3. Ishii, M., Sukanya, P., & Sato, M. (1994). A virtual work space for both hands manipulation with coherency between kinesthetic and visual sensation. In *Proceedings of the forth international symposium on measurement and control in robotics*, December 1994 (pp. 84–90).
4. Cai, Y., Ishii, M., & Sato, M. (1996). A human interface device for CAVE size virtual space. In *Proceedings of the IEEE international conference on systems, man and cybernetics*, October 1996 (pp. 2084–2089).
5. Kim, S., Ishii, M., Koike, Y., & Sato, M. (2000). Design a tension based haptic interface: SPIDAR-G. In *World multiconference on systemics* (pp. 422–427).
6. Yanlin, L., Murayama, J., Akahane, K., Hasegawa, S., & Sato, M. (2003). Development of new force feedback interface for two-handed 6DOF manipulation SPIDAR-G&G system. In *ICAT 2003* (pp. 166–172).
7. Zhu, Y., Koyama, T., Igarashi, T., Akahane, K., & Sato, M. (2011). Development of desktop haptic interface SPIDAR-I. In *ASIAGRAPH2011*, 20–22 October 2011 (Vol. 5, No. 1, pp. 26–29).
8. Akahane, K., Hasegawa, S., Koike, Y., & Sato, M. (2006). A proposal of a high definition haptic rendering for stability and fidelity. In *ICAT*.
9. Hasegawa, S., Ishii, M., Sato, M., & Koike, Y. (1999). Inter-process communication for force display of dynamic virtual world. In *Proceedings of the ASME dynamic systems and control division*, November 1999 (Vol. 67, pp. 211–218).
10. Kawamura, S., & Ito, K. (1993). A new type of master robot for teleoperation using a radial wire drive system. In *Proceedings of the IEEE/RSJ IROS* (pp. 55–60).
11. Corlett, N. (1986). *The ergonomics of working postures*. London: Taylor & Francis.
12. Walairacht, S., Koike, Y., & Sato, M. (1999). A new haptic display for both-hands-operation: SPIDAR-8. In *Proceedings of the IEEE international symposium on intelligent signal processing and communication systems*, December 1999 (pp. 569–572).
13. Walairacht, S., Koike, Y., & Sato, M. (2000). String-based haptic interface device for multi-fingers. In *Proceedings of the IEEE virtual reality 2000*, March 2000 (p. 293).
14. Walairacht, S., Yamada, K., Koike, Y., & Sato, M. (1999). Modeling virtual hands with haptic interface device. In *Proceedings of the ninth international conference on artificial reality and tele-existence*, December 1999 (pp. 233–236).
15. Hyun, J., Kumazawa, I., & Sato, M. (2011). Controllability evaluation for handheld devices using AR techniques: pre-evaluation of button characteristics using the SPIDAR-hand system. *Transactions of the Virtual Reality Society of Japan*, *16*(2), 189–200. Special issue on mixed reality 5.

Printed in the United States
By Bookmasters